D0142653

Rotary, Kelly, Swivel, Tongs, and Top Drive

Unit I, Lesson 4
First Edition

Formerly *Rotary, Kelly, and Swivel*

By L. D. Davis

Published by

PETROLEUM EXTENSION SERVICE
Division of Continuing Education
The University of Texas at Austin
Austin, Texas

in cooperation with

INTERNATIONAL ASSOCIATION
OF DRILLING CONTRACTORS
Houston, Texas

1995

Library of Congress Cataloging-in-Publication Data

Davis, L.D. 1953—
Rotary, kelly, swivel, tongs, and top drive / L.D. Davis. — 1st ed.
p. cm. — (Rotary drilling series ; unit I, lesson 4)
"Published. . . . in cooperation with International Association of Drilling Contractors."
"Catalog no. 2.104101"—T.p. verso.
Includes index.
ISBN 0-88698-172-7 (pbk.)
1. Oil well drilling rigs—Equipment and supplies. I. University of Texas at Austin. Petroleum Extension Service. II. International Association of Drilling Contractors. III. Title. IV. Series.
TN871.5.D335 1995
622'.3381—dc20

95-45843

CIP

First Edition published 1995. Third Impression 2001
Printed in the United States of America

Catalog no. 2.104101
ISBN 0-88698-172-7

Contents

Figures

Table

Foreword

▼
▼
▼

For many years, the Rotary Drilling Series has oriented new personnel and further assisted experienced hands in the rotary drilling industry. As the industry changes, so must the manuals in this series reflect those changes.

The revisions to both text and illustrations are extensive. In addition, the layout has been "modernized" to make the information easy to get; the study questions have been rewritten; and each major section has been summarized to provide a handy comprehension check for the student.

PETEX wishes to thank industry reviewers—and our readers—for invaluable assistance in the revision of the Rotary Drilling Series. On the PETEX staff, Deborah Caples designed the layout; Doris Dickey proofread innumerable versions; and Sheryl Horton saw production through from idea to book.

Although every effort was made to ensure accuracy, this manual is intended to be only a training aid; thus, nothing in it should be construed as approval or disapproval of any specific product or practice.

Ron Baker

Acknowledgments

Special thanks to Ken Fischer, director, Committee Operations, International Association of Drilling Contractors, who reviewed this manual and secured other reviewers. John Altermann, Reading & Bates Drilling Company, Joey Hopewell, Rowan Petroleum Incorporated, and Jim Arnold, Salem Investment Corporation, provided invaluable suggestions on the content and language. Without their assistance, this book could not have been written. In addition, special thanks to Leslie Kell, who managed to interpret some difficult sketches and make them into excellent drawings, and to Terry Gregston, for her excellent black and white photographs.

Units of Measurement

▼
▼
▼

Throughout the world, two systems of measurement dominate: the English system and the metric system. Today, the United States is almost the only country that employs the English system.

The English system uses the pound as the unit of weight, the foot as the unit of length, and the gallon as the unit of capacity. In the English system, for example, 1 foot equals 12 inches, 1 yard equals 36 inches, and 1 mile equals 5,280 feet or 1,760 yards.

The metric system uses the gram as the unit of weight, the metre as the unit of length, and the litre as the unit of capacity. In the metric system, for example, 1 metre equals 10 decimetres, 100 centimetres, or 1,000 millimetres. A kilometre equals 1,000 metres. The metric system, unlike the English system, uses a base of 10; thus, it is easy to convert from one unit to another. To convert from one unit to another in the English system, you must memorize or look up the values.

In the late 1970s, the Eleventh General Conference on Weights and Measures described and adopted the Système International (SI) d'Unités. Conference participants based the SI system on the metric system and designed it as an international standard of measurement.

The *Rotary Drilling Series* gives both English and SI units. And because the SI system employs the British spelling of many of the terms, the book follows those spelling rules as well. The unit of length, for example, is *metre*, not *meter*. (Note, however, that the unit of weight is *gram*, not *gramme*.)

To aid US readers in making and understanding the conversion to the SI system, we include the following table.

Metric Conversion Factors

Quantity or Property	English Units	Multiply English Units By	To Obtain These SI Units
Length, depth, or height	inches (in.)	25.4	millimetres (mm)
		2.54	centimetres (cm)
	feet (ft)	0.3048	metres (m)
	yards (yd)	0.9144	metres (m)
	miles (mi)	1609.344	metres (m)
		1.61	kilometres (km)
Hole and pipe diameters, bit size	inches (in.)	25.4	millimetres (mm)
Drilling rate	feet per hour (ft/h)	0.3048	metres per hour (m/h)
Weight on bit	pounds (lb)	0.445	decanewtons (dN)
Nozzle size	32nds of an inch	0.794	millimetres (mm)
Volume	barrels (bbl)	0.159	cubic metres (m^3)
		159	litres (L)
	gallons per stroke (gal/stroke)	0.00379	cubic metres per stroke (m^3/stroke)
	ounces (oz)	29.57	millilitres (mL)
	cubic inches ($in.^3$)	16.387	cubic centimetres (cm^3)
	cubic feet (ft^3)	28.3169	litres (L)
		0.0283	cubic metres (m^3)
	quarts (qt)	0.9464	litres (L)
	gallons (gal)	3.7854	litres (L)
	gallons (gal)	0.00379	cubic metres (m^3)
Pump output and flow rate	gallons per minute (gpm)	0.00379	cubic metres per minute (m^3/min)
	gallons per hour (gph)	0.00379	cubic metres per hour (m^3/h)
	barrels per stroke (bbl/stroke)	0.159	cubic metres per stroke (m^3/stroke)
	barrels per minute (bbl/min)	0.159	cubic metres per minute (m^3/min)
Pressure	pounds per square inch (psi)	6.895	kilopascals (kPa)
		0.006895	megapascals (MPa)
Temperature	°Fahrenheit (°F)	$\frac{°F - 32}{1.8}$	°Celsius (°C)
Mud weight	pounds per gallon (ppg)	119.82	kilograms per cubic metre (kg/m^3)
	pounds per cubic foot (lb/ft^3)	16.0	kilograms per cubic metre (kg/m^3)
Mass (weight)	ounces (oz)	28.35	grams (g)
	pounds (lb)	453.59	grams (g)
		0.4536	kilograms (kg)
	tons (tn)	0.9072	tonnes (t)
Pressure gradient	pounds per square inch per foot (psi/ft)	22.621	kilopascals per metre (kPa/m)
Funnel viscosity	seconds per quart (s/qt)	1.057	seconds per litre (s/L)
Power	horsepower (hp)	0.7	kilowatts (kW)
Area	square inches ($in.^2$)	6.45	square centimetres (cm^2)
	square feet (ft^2)	0.0929	square metres (m^2)
	square yards (yd^2)	0.8361	square metres (m^2)
	square miles (mi^2)	2.59	square kilometres (km^2)
	acre (ac)	0.40	hectare (ha)
Drilling line wear	ton-miles (tn•mi)	14.317	megajoules (MJ)
		1.459	tonne-kilometres (t•km)
Torque	foot-pounds (ft•lb)	1.3558	newton metres (N•m)

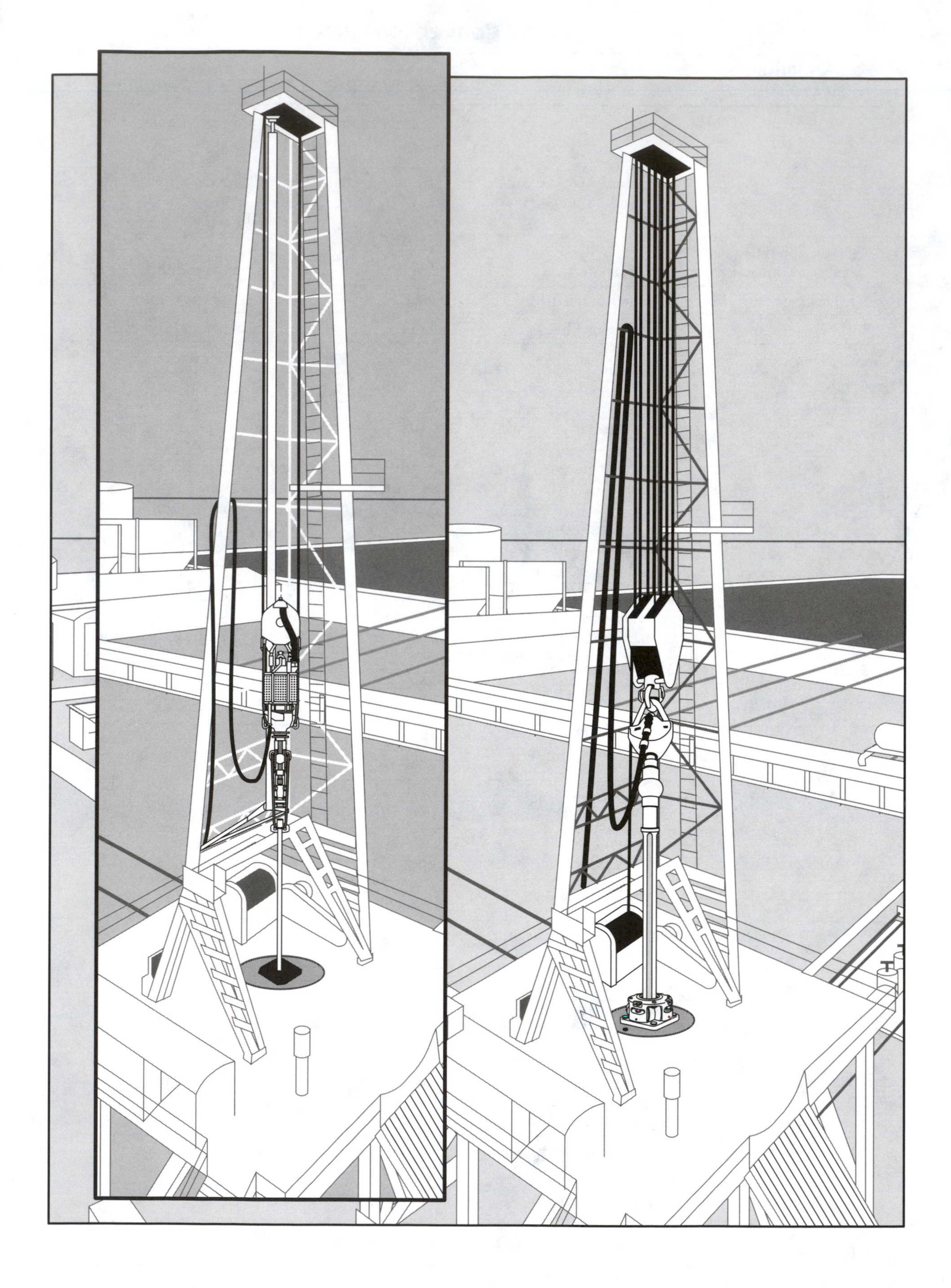

Introduction

▼
▼
▼

Drillers ready to drill ahead sometimes say, "Let's put the bit on bottom and turn it to the right." This oil patch expression is a nod to a special technology called rotary drilling. Rotary drilling bores through underground formations by rotating (turning) the drill stem and the bit.

Today, rotary drilling is the industry standard, but it was not always so. Before rotary drilling started to flourish in Texas in the 1900s, oil people drilled most wells with cable drilling tools. With this method, rig crew members attach a sharp tool—a bit—to a cable. The cable, along with other rig equipment, repeatedly picks up and drops the heavily weighted bit, which punches a hole into the ground.

Cable tool drilling has two big drawbacks. Chips of rock (cuttings) that the bit gouges from the formation stay in the hole. The cable-tool system has no way of getting them out of the way as the bit drills. Eventually, the cuttings build up to the point that the bit starts punching into old cuttings instead of into fresh, uncut rock. At this point, the bit no longer deepens the hole. Crew members therefore have to stop the operation and bail (remove) the cuttings. Even worse, however, is that some soft formations cave in around the bit and keep it from drilling at all.

Rotary drilling solves the problems of having to stop drilling to bail, and of cave-ins in soft formations. The beauty of the rotary method is that it not only rotates the bit to drill ahead (make hole), it also removes cuttings from the hole at the same time. Removing cuttings at the same time the bit drills keeps the hole clean, regardless of how soft the formations are. Unlike cable-tool drilling, rotary drilling uses hollow pipe (the drill stem) to put the bit on the bottom of the hole. The diameter of the bit is larger than the diameter of the drill stem, so it drills a hole whose diameter is larger than the drill stem's. Thus, there is space between the drill stem and the wall of the hole. This space is the annular space, or the annulus.

On a rotary rig, a powerful pump circulates fluid (drilling mud) down the drill stem and out of the bit at the same time the bit drills. The circulating fluid lifts the cuttings up the annular space to the surface. There, special machines remove the cuttings from the fluid before the pump recirculates the clean fluid back down the drill stem. The pump constantly circulates drilling fluid whenever the bit is on bottom and drilling.

Surface equipment called the rotating system makes rotary drilling possible. Rigs use one of two rotating systems: a conventional rotary table system or a top-drive system. The conventional rotary table system has been around, with improvements, of course, since rotary rigs began to dominate drilling in the early-to-middle part of the twentieth century.

A relatively new rotary drilling method uses a top drive. Manufacturers developed top drives in the 1970s and rig owners began using them in a big way in the late 1980s. Top drives are popular because, even though they are expensive, they are more efficient to use than the conventional rotary table system. A top-drive system is a kind of power swivel. It hangs from the hook like a regular swivel, but, unlike a regular swivel, it has a very powerful motor that has a threaded drive shaft. Crew members make up the drill stem to the shaft and the motor turns the drill stem and bit.

▼
▼
▼

Conventional Rotating System

▼
▼
▼

Figure 1 shows a conventional rotary system. From top to bottom, it consists of a hook, a swivel, and a rotary (kelly) hose. It also has an upper kelly cock (valve), a kelly, and a lower kelly cock (valve), which is screwed into the bottom of the kelly and cannot be seen on the figure. Not shown in the figure, but an important part (you will see why later), is a kelly saver sub.

The conventional rotating system also has a kelly bushing, a master bushing, and a rotary table assembly. The rotating system allows part of the drill stem's weight to press down on the bit to make it drill. The system also provides the rotating force to turn the bit. Finally, it provides a passageway for the pump to send drilling fluid downhole to lift cuttings.

Let's take a closer look at how a conventional rotary rig accomplishes these three jobs (fig. 2). The drilling crew attaches a drill bit studded with metal or diamond cutters to the bottom of the drill stem. Crew members then lower the drill stem into the hole until the bit is very near the bottom. At this point, the driller engages the rotary table assembly on the rig floor to turn the drill stem and bit. The mud pump is also started to circulate drilling fluid. The driller then lowers the rotating bit the rest of the way to bottom and allows part of the drill stem's weight to push down on the bit. Weight causes the bit's cutters to bite into the formation and drill ahead.

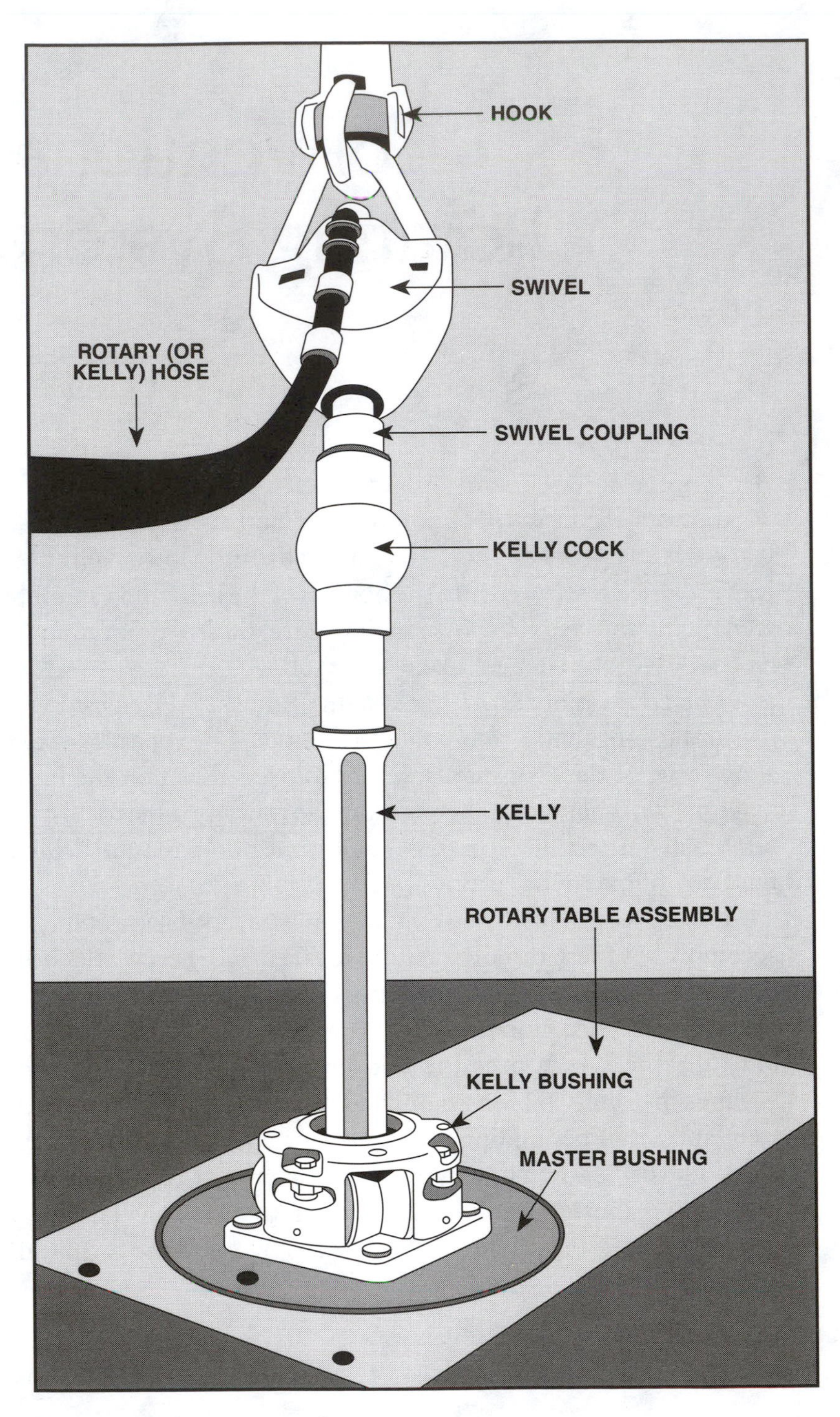

Figure 1. A conventional rotary system

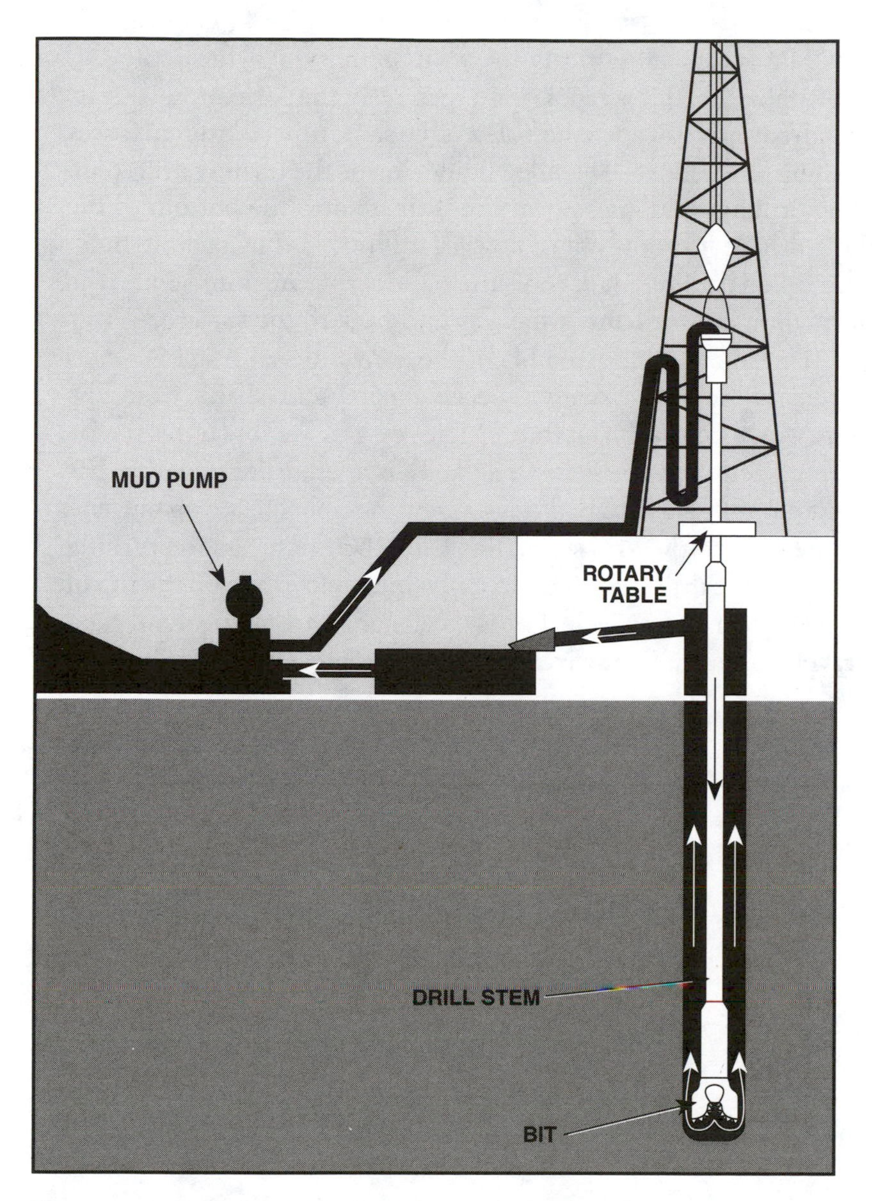

Figure 2. As the bit drills, drilling fluid circulates.

The rotary table turns the drill stem, which turns the bit. Downhole, the bit scrapes or gouges away the formation with its tough cutters to deepen the hole. At the same time, the pump forces drilling fluid, a special blended liquid, inside the turning drill stem. The drilling fluid goes down the drill stem to the bottom of the hole, where it shoots out of nozzles on the bit. The drilling fluid, also called drilling mud, cools the rapidly rotating bit and lifts the formation cuttings out of the way and back to the surface.

The rotary table assembly of the conventional system creates a turning action. Usually, special chains and shafts from the drawworks power the assembly. The rotary assembly transfers the turning action to the kelly. The kelly is a length of pipe that has either four or six flattened sides. Crew members make up one end of the kelly to the drill stem. The kelly then transfers the turning action to the drill stem and bit. Crew members make up the top of the kelly to the swivel. The swivel hangs from the hook and traveling block. The swivel supports the kelly and the drill string and allows the kelly to spin freely.

▼
▼
▼

Top-Drive System

▼
▼
▼

Some rigs, especially those offshore, do not use a rotary table, a kelly, and a swivel to rotate the drill stem. Instead, they use a top drive, which replaces the rotary table, the kelly, and the swivel. The top drive does the work of all three and works much like a motorized swivel (fig. 3). Because of its design, the top drive can speed up the rotary drilling process. Like the swivel, it hangs from the hook on the traveling block. Unlike the swivel, it has a heavy-duty motor (or motors) that provides power to rotate the drill stem.

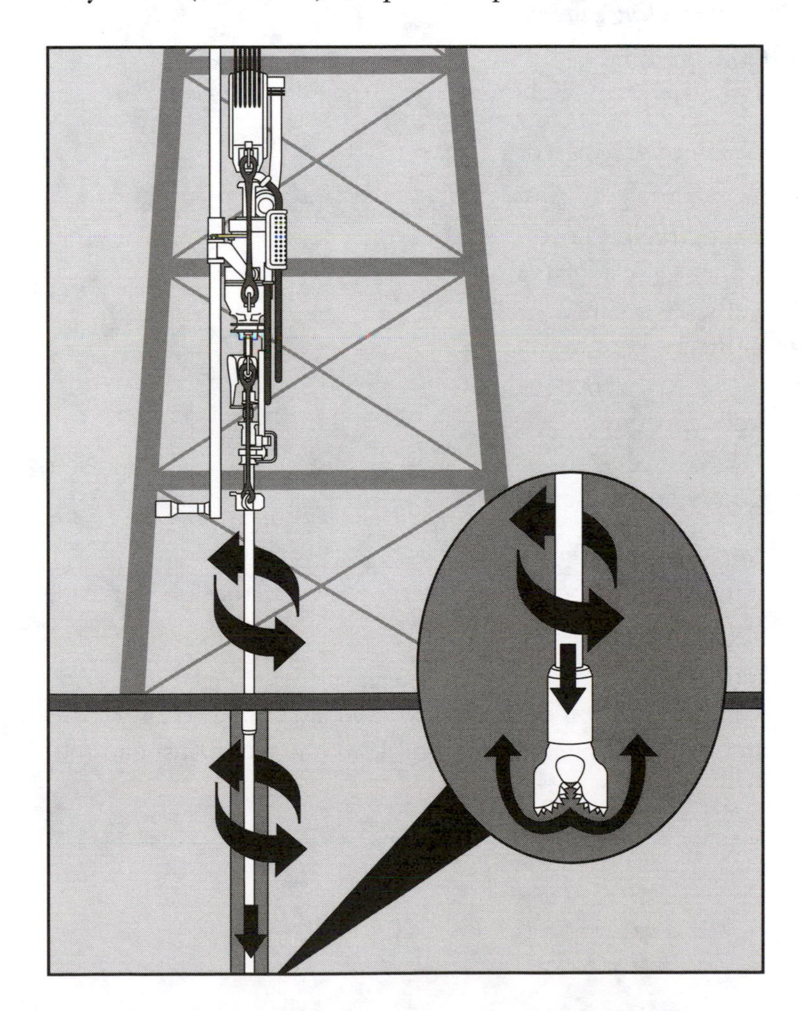

Figure 3. A top-drive system does not use the rotary table to turn the drill stem.

The uppermost stand of drill pipe threads into a drive shaft on the top drive. The top drive rotates the entire drill stem and bit directly. This action eliminates the need for a turning rotary table assembly and a kelly. Crew members use the rotary table only as a place to suspend the drill string when adding pipe or tripping in and out.

To summarize—

Three processes of rotary drilling

- Downward movement of the drill stem
- Turning the bit
- Removing the rock cuttings from the hole by circulating drilling fluid

Two types of rotating systems

- Conventional
- Top drive

Conventional rotating parts

- Swivel
- Rotary (kelly) hose
- Upper kelly cock (valve)
- Kelly saver sub
- Kelly
- Lower kelly cock (valve)
- Kelly bushing
- Master bushing
- Rotary table assembly

The top-drive system eliminates

- Kelly
- Kelly bushing
- Kelly cocks
- Rotating portion of the rotary table assembly (crew members still need a rotary table as a place to suspend the drill stem)

▼
▼
▼

Rotary Table Assembly

▼
▼
▼

Definition

A conventional rotary rig uses a conventional rotary table assembly (fig. 4). This assembly is a rotating machine housed inside a rectangular steel box. The assembly has an opening in the middle for the kelly and drill pipe. The main parts of the rotary table assembly include the base, the turntable (also called the rotary table), and the master bushing. It also has a drive-shaft assembly, a drawworks sprocket and a drive-shaft sprocket, and a chain.

Righands often call the steel box and the equipment it houses the rotary, or the rotary table. Strictly speaking, however, a rotary table is a collection of many components. One of its main parts is a turning component called the rotary table, or the turntable. This manual will refer to the entire rotary machine as the rotary table assembly and refer to the turning device in the assembly as the rotary table.

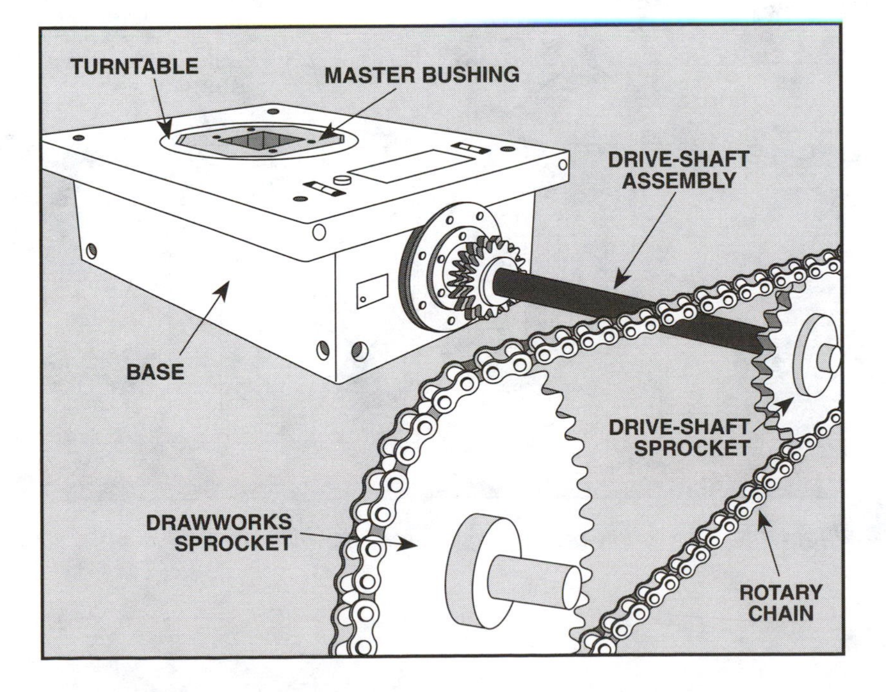

Figure 4. A conventional rotary table assembly

Functions

The rotary table assembly has two main functions:—

1. During drilling, it rotates and transfers the turning motion to the kelly.
2. When drilling is stopped, it holds (suspends) the weight of the drill string when slips are set in the rotary table.

During Drilling

During normal drilling, the driller operates the rotary table assembly from the driller's console. From the console (the operating center), the driller controls the direction and speed of rotation. Once the bit is on bottom, the driller turns the rotary to the right, or clockwise. The driller can also adjust the rotary table to turn from about 20 to over 200 revolutions per minute (rpm), depending on the type and size of the bit and other factors.

When Drilling Stops

When drilling stops, the crew uses slips (steel wedges with gripping elements) to hold the drill stem in the rotary table. The rotary table assembly and the slips act like a vice to hold the top of the drill stem (fig. 5). With the drill stem suspended in the rotary table assembly, the crew can make up or break out drill pipe or other drill stem devices.

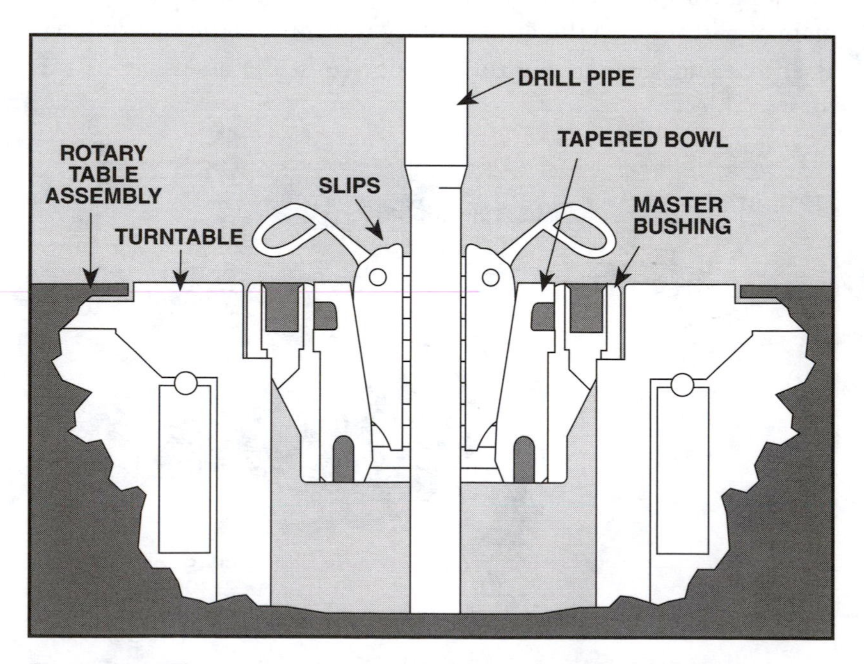

Figure 5. The rotary table assembly and slips hold the drill stem.

The driller uses the drawworks to lift the bit off bottom, and the crew sets the slips between the rotary table and the pipe. When placed inside the rotary table, slips grip the pipe and suspend it in the hole. The rotary table then supports the weight of the drill stem.

While the slips suspend the pipe in the rotary table, the driller can also use the assembly to spin out a pipe connection. The rotary helpers first loosen the connection with the breakout tongs and the backup tongs. Tongs are large wrenches the crew latches onto the pipe. If the rig does not have a spinning wrench (a wrench that rapidly unthreads the joint), the driller turns the rotary table counterclockwise to spin out the joint.

Keep in mind, however, that the entire string below the rotary rotates when spun out with the rotary. Thus, if anything is in the string that could come unscrewed as a result of this spinning motion, the driller should not use the rotary to spin out a joint.

It is also important that the driller not use the rotary table to spin up or tighten a joint. It is difficult to control the amount of torque (twisting force) the table puts on the pipe as it spins. Therefore, using the rotary table to spin up, or tighten, tool joints can easily damage the pipe by over- or undertorquing it.

How the Rotary Table Assembly Works

Usually, two sprockets connected by a large chain drive the rotary table assembly (fig. 6). The rotary table assembly drive works the same way as two sprockets and chain transfer power on a bicycle. When the driller engages the rotary table assembly, power from the drawworks turns the drawworks sprocket. The drawworks sprocket sends power through the rotary chain to the drive-shaft sprocket. This second sprocket transfers power to the drive-shaft assembly. The other end of the drive shaft has a pinion, which is a gear with beveled teeth. The pinion meshes with a circular gear, the ring gear, on the rotary table itself, causing it to turn.

Some rigs have independent rotary drive systems. In an independent drive, the rig builder mounts the rotary table and a power source on a single skid. The power source is usually a small diesel engine or an electric motor. If the power source is an engine, a gear box and a fluid coupling transfer the engine's power to the rotary table assembly.

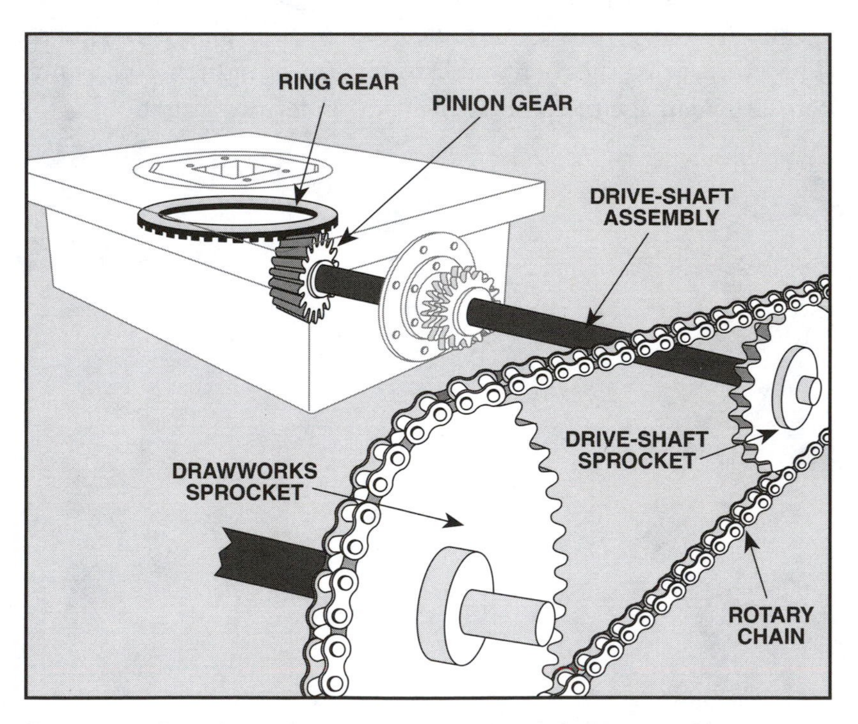

Figure 6. Sprockets, chains, and gears drive the rotary table.

Size

Rig owners size rotary table assemblies by the diameter of the opening in the middle where the crew suspends the drill stem. Typical sizes range from 17½ inches (445 millimetres) to 49½ inches (1,257 millimetres), with three sizes in between. The American Petroleum Institute (API) sets the standards for rotary tables as well as other oilfield equipment. Equipment built to API standards ensures that a rig owner can buy it anywhere in the world and feel confident that it will fit and work on any rig.

Rig owners must also know the rotary table assembly's load capacity. Load capacity indicates how much weight a rotary table assembly can safely bear. Load capacities range from 100 tons (91 tonnes) to 600 tons (546 tonnes). Generally, the larger the rig, the larger the rotary table assembly is and the greater its load capacity.

Components

Base

The base (fig. 7) is a cast steel or reinforced steel shell that encloses the pinion end of the drive shaft and the rotary table. The boxlike base serves several functions. First, it makes the rotary equipment a single, portable unit. As such, crew members can easily move and install it. Second, the base protects the rotating machinery inside from drilling mud and the elements. Finally, the base bears the weight of the drill stem and other downhole tools.

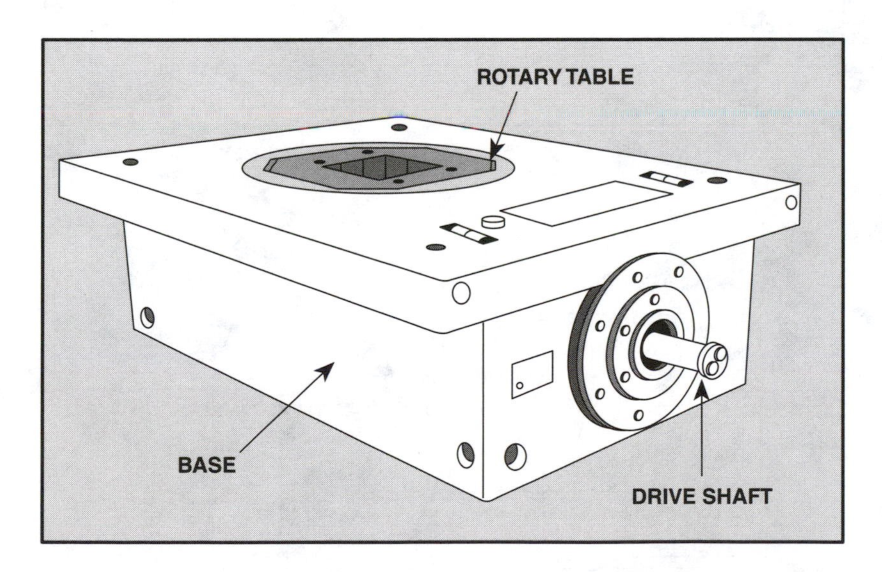

Figure 7. Rotary table base

Inside (fig. 8) the manufacturer mounts the turntable (the rotary table) securely on a frame. A lock assembly allows crew members to keep the rotary table from turning. An oil bath reservoir, in which some of the rotary table's components sit, is part of the lubrication system. The manufacturer provides an oil gauge (a dipstick) to allow the crew to check the reservoir's oil level. The manufacturer also provides grease fittings. Crew members use a grease gun on these fittings at regular times to keep components not in the oil bath lubricated.

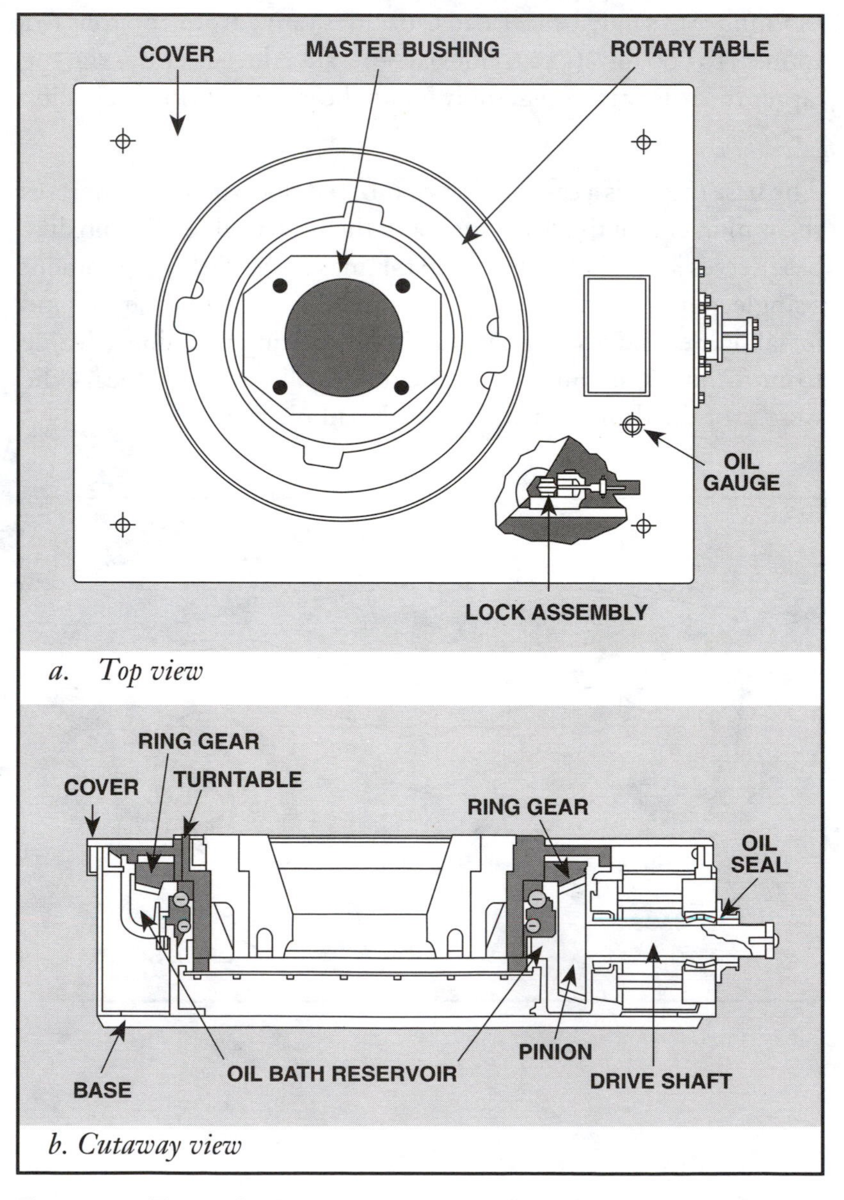

Figure 8. Top and side (cutaway) views of rotary table assembly

Rotary Table

The rotary table is also called the rotary, the turntable, or, simply, the table (fig. 9). It is the main piece of equipment in the assembly. The rotary table consists of circular gears, seals, and ball bearings. These items make up a turning device that has an opening in the middle for drill pipe. The manufacturer locks the rotary table into the middle of the base.

The function of the rotary table is twofold. First, it transmits a rotating motion to a bushing (a special steel cylinder), which rotates the kelly. The kelly, attached to the topmost drill pipe joint, then turns the drill stem. Second, when the crew suspends the drill stem in the hole, the rotary table also supports the weight of the entire drill stem.

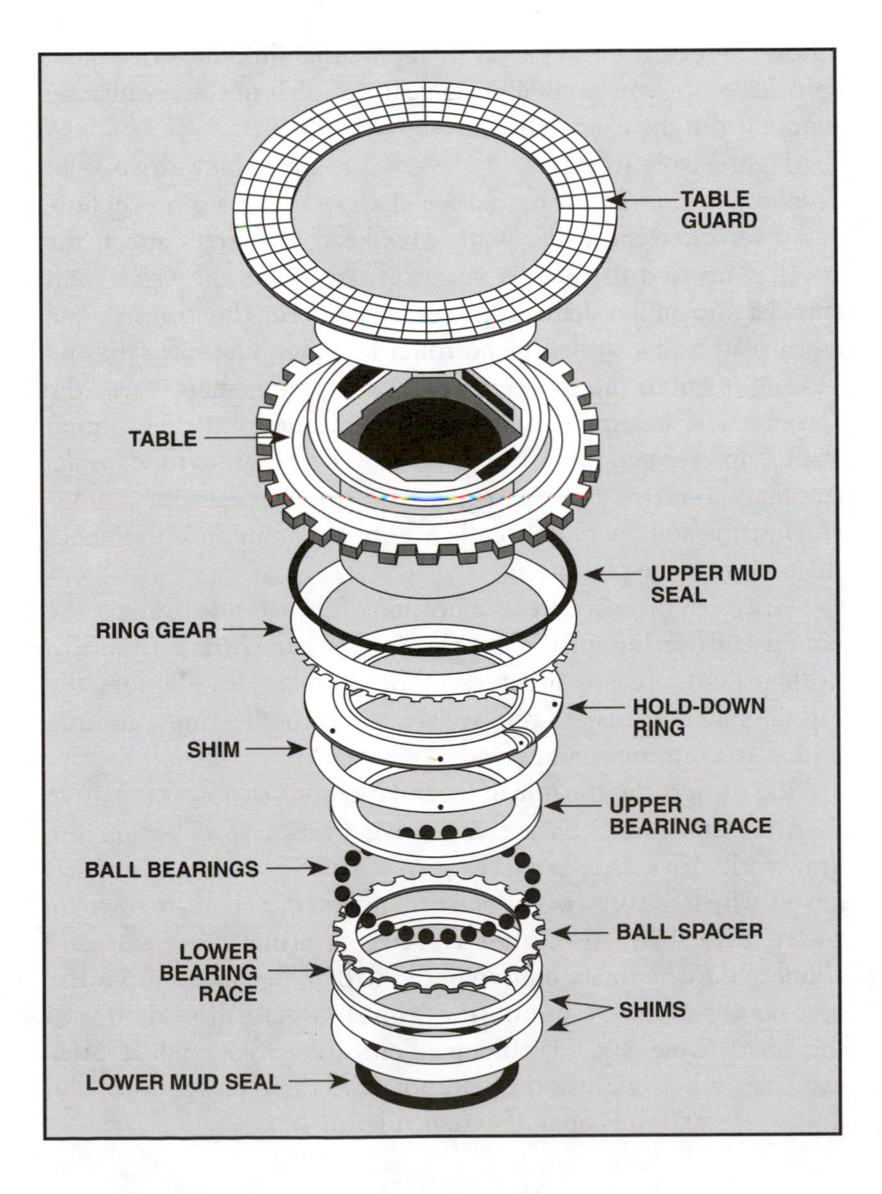

Figure 9. Exploded view of rotary table

Primary components of the rotary table include the table, the ring gear, the ball bearings and races, and the upper and lower mud seals. Notches on the rim of the table are part of a locking device that holds the table stationary. When a crew member inserts a locking bolt between the notches, it prevents the table from turning. The ring gear interfaces with the pinion gear on the drive shaft (see fig. 6). They turn the rotary table. Ball bearings usually make up the main bearing. A few rotary tables, however, use roller bearings. The upper and lower mud seals protect the moving gears from drilling mud. Manufacturers make them out of neoprene (a synthetic rubber). Springs hold them in place. In most rotaries, the manufacturer builds in the upper mud seal, so crew members do not have to service it.

Drive-Shaft Assembly

The drive shaft is a heavy-duty metal rod that links the drive-shaft sprocket to the rotary table's ring gear. Its job is to convey turning motion from the rotary chain to the rotary table.

Figure 10 is an exploded view of a rotary table's drive-shaft assembly. The assembly has a drive-shaft sprocket, a sprocket hub, and a sprocket-end cover plate. Hex head capscrews attach the cover plate to a capsule. A sprocket key holds the drive-shaft sprocket to a pinion shaft. An oil seal fits between the sprocket-end cover plate and a sprocket-end roller bearing. The sprocket-end bearing fits onto the sprocket end of the pinion shaft where the sprocket-end bearing's inner race is machined onto the pinion shaft. Capscrews connect the pinion-end cover plate to the capsule. Another oil seal (not visible in the figure) goes between the pinion-end bearing and the pinion-end cover plate. A pinion key attaches the pinion to the pinion shaft.

Grease or oil lubricates the bearings. The capsule protects the pinion-end bearing and the sprocket-end bearing from exposure to drilling fluids. It also provides lubricant seals. In addition, the capsule makes it easy to remove the shaft, the bearings, and the pinion as a unit for maintenance.

Rig owners can use one of several methods to power the drive shaft. Most commonly, a chain drive from a sprocket on the drawworks drives the shaft (see fig. 6). A few rigs use another shaft, or rod, which the drawworks powers, instead of a chain, to turn the rotary drive shaft. Also, some rigs have a prime mover directly connected to the rotary unit. In one kind of independent drive, the base houses an electric motor (the prime mover), a drive shaft, and the rotary table (fig. 11). Independent-drive rotary tables offer extra flexibility because they do not take their power from the drawworks and thus operate independently of it.

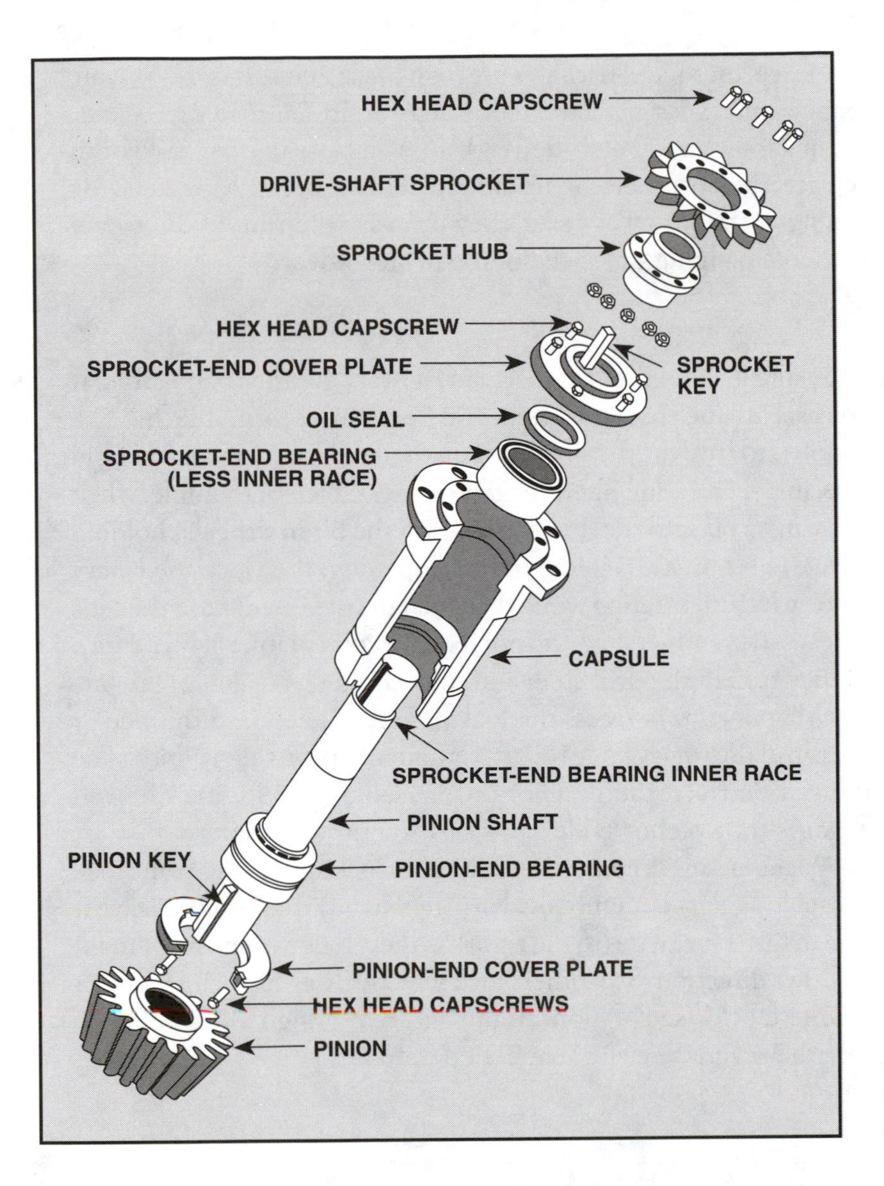

Figure 10. Exploded view of a rotary table drive shaft assembly

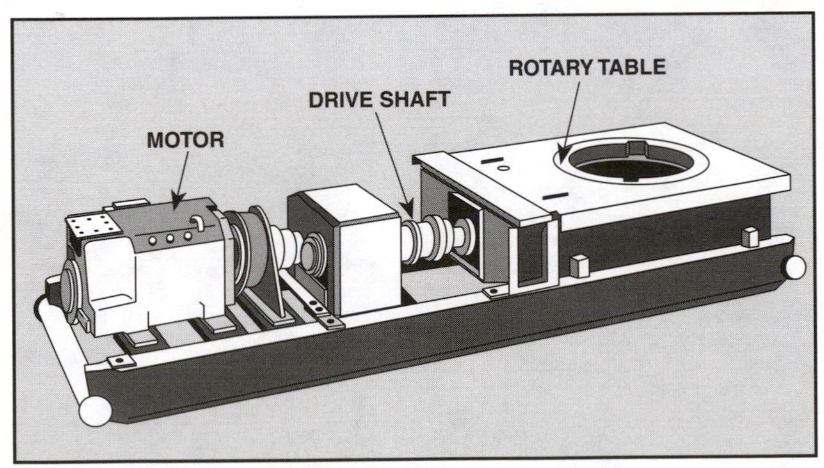

Figure 11. An electric motor drives this rotary table assembly.

Sprockets

Sprockets on rotary machines are high-wear items. For this reason, manufacturers design them to be detachable. In detachable sprockets, the manufacturer bolts the teeth to a hub so that righands can replace the sprockets or install smaller or larger sprockets. By changing the sprocket's size, they can vary the rpm of the rotary without changing the speed of the prime mover.

Locking Devices

A locking device is a small mechanical brake for the rotary table. It consists of a lock pin and notches on the rotary's turntable (fig. 12). Its job is to stop the turning movement of the rotary table and hold it securely. Crew members lock the rotary table, for example, when screwing a bit into the drill stem. With the bit in a special holding device (a bit breaker) in the rotary's opening, they lock the rotary to keep it from rotating while they turn the drill stem onto the bit's threads. If crew members did not lock the rotary table, it would turn as they turned the drill stem and they could not make up the bit.

Manufacturers recess the locking pin or latch into the side or the top of the rotary table base. Crew members on the rig floor slide the rod between the notches on the table (see fig. 9). The rod engages the notched table and stops any turning motion.

Manufacturers make several kinds of locking device. Some lock the table against counterclockwise (left-hand) rotation but leave it free to turn clockwise (to the right). Others lock against movement in either direction. Still others lock selectively against clockwise or counterclockwise rotation. Manufacturers design all of them so that they cannot be accidentally operated.

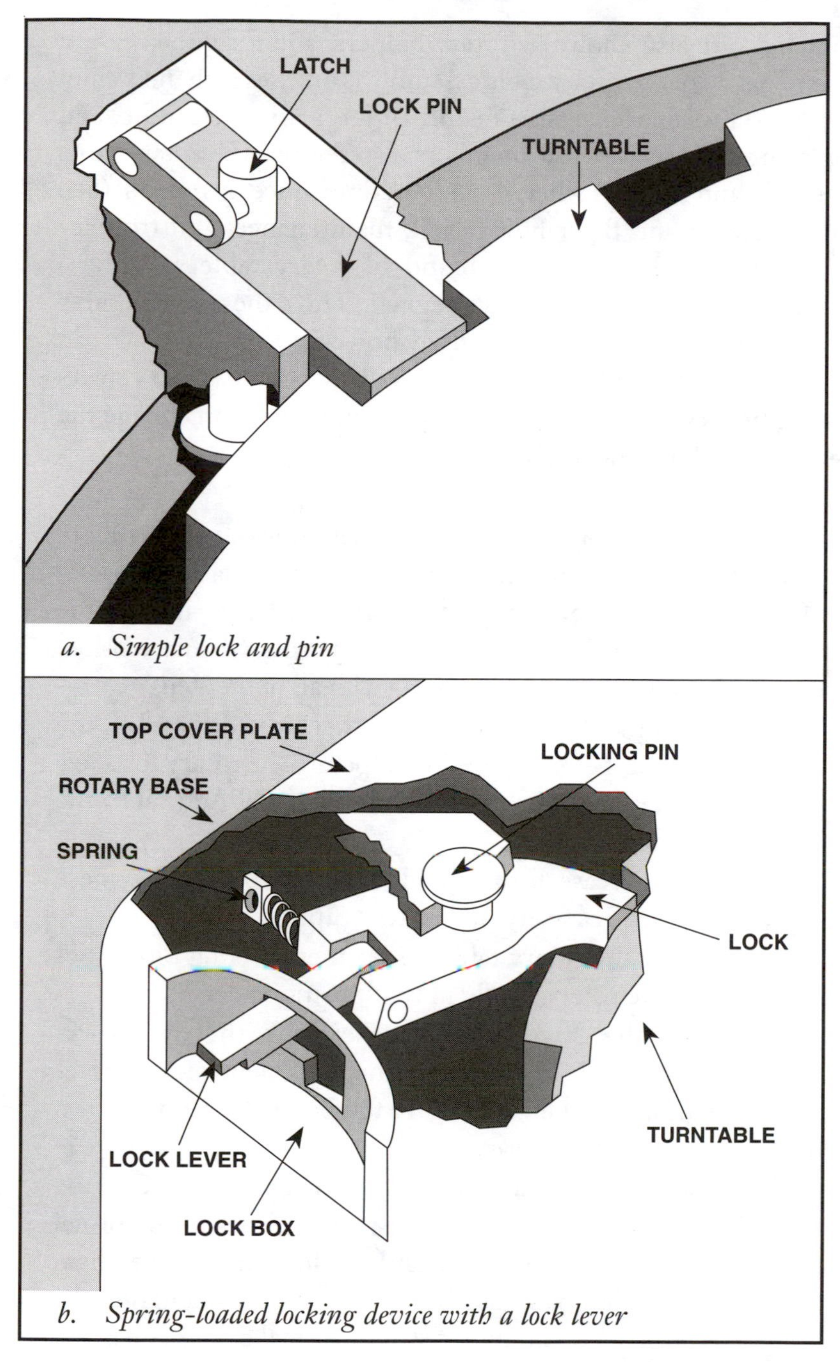

Figure 12. Rotary locking devices

Maintenance

Righands are also known as rotary helpers, and for good reason: maintenance of the rotary table assembly is one of a righand's chief duties. Maintenance consists mainly of lubrication, cleaning, and replacing mud seals. Crew members also remove the rotary table assembly and install it when it needs maintenance or repairs. (See Appendixes A and B for lubrication, maintenance, and trouble-shooting procedures for one brand of rotary table assembly.) Rotary table assemblies are very rugged. Thus, the manufacturer seldom has to repair them. If required, however, rig supervisors can plan for repair in advance. They usually schedule these repairs during times when the rig is idle so crew members can change the assembly with no added downtime.

Installation

Righands install the rotary assembly when it returns from the shop, or when the rig owner buys a new one. Correct installation involves five main points: (1) oil level, (2) grease lubrication, (3) positioning the rotary table, (4) sprocket alignment, and (5) rotary chain tension. Crew members should become familiar with all of the steps.

1. Using the dipstick (fig. 13a), check the oil level in the reservoir before operating. Operate the rotary for a few minutes and then check the level again. Add oil to the specified level.
2. Check grease fittings and be sure to lubricate these points if necessary before start-up (fig. 13a).
3. Center the rotary table under the traveling block. Secure the rotary firmly in that position.
4. Align the drive-shaft sprocket with the drawworks sprocket (fig. 13b). Adjust the alignment of the drive-shaft sprocket by moving it a few inches (centimetres) on the shaft if necessary.
5. Check the tension on the drive chain between the drawworks sprocket and the rotary sprocket. In normal operation, the lower strand of the rotary drive chain carries the load, and the upper strand maintains the slack. Check the upper strand to make sure it has the proper slack. To check the slack, lock the rotary table assembly and pull the lower chain taut. There should be enough slack in the upper chain to allow it to drop 2 to 3 inches (50 to 75 millimetres) below a straight line at its midpoint (fig. 13c).

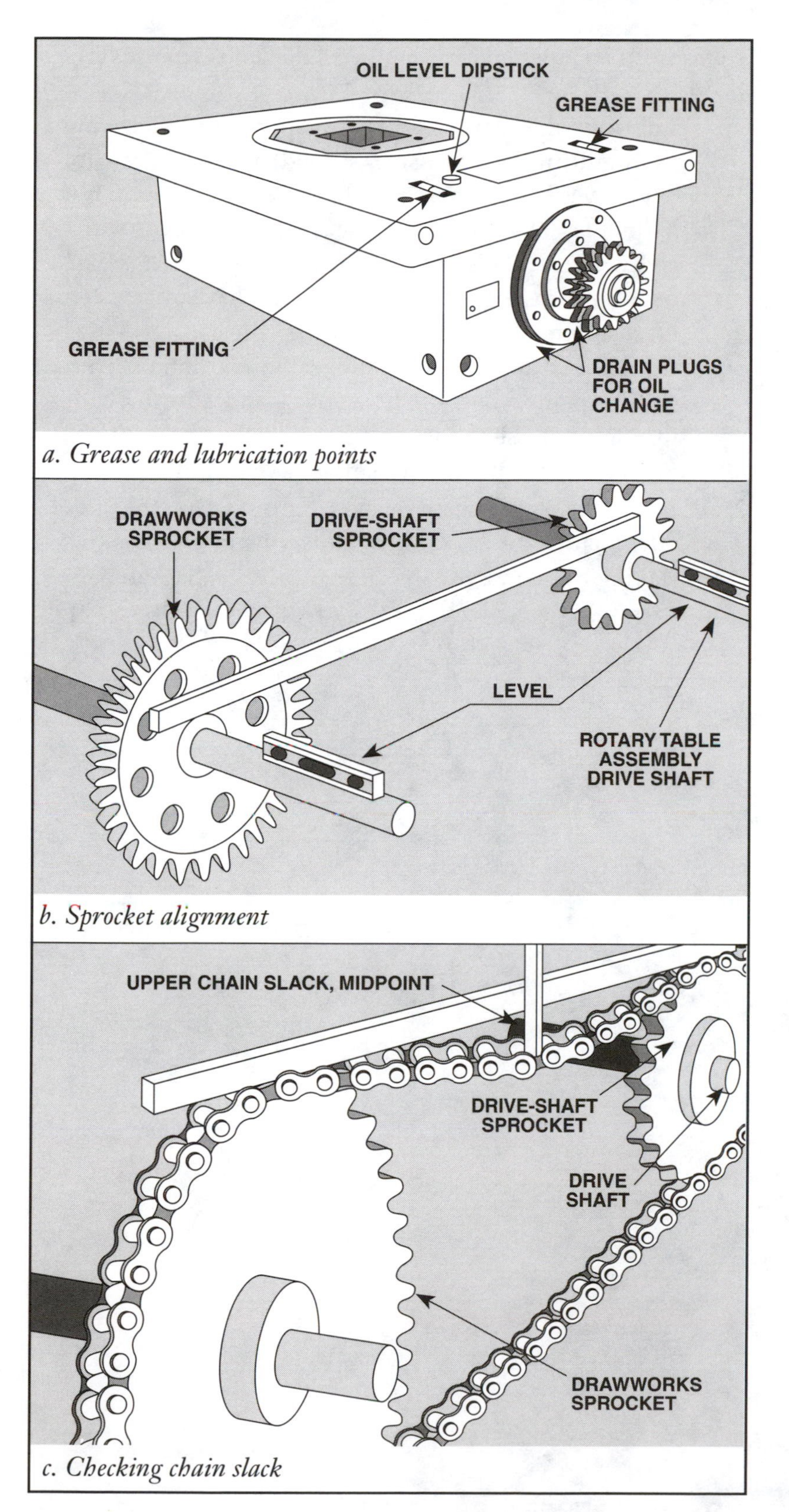

Figure 13. Rotary table assembly maintenance points

Installation also includes lubricating the chain and then installing the guard.

1. Lubricate the chain before start-up. Many contractors use an oil-tight case (a steel shroud) around the chain. The lower end of this shroud has a sump. A sump is a basin that holds oil, and the chain runs through it continuously. This type of shroud not only is a guard, but also keeps the oil inside. Some rigs have a force-fed lubricating system for the rotary chain. It also uses a steel shroud and an oil bath. Unlike the continuous-bath type, a pump takes oil from the sump and forces it through a pressure line. Oil then drips from this perforated line onto the chain's surfaces.
2. Install the rotary chain guard before start-up (fig. 14). The guard protects crew members from harm, so installing it is an essential step in chain installation.

Figure 14. Rotary drive chain guard

Lubrication

Lubrication is the key part of rotary table assembly maintenance. The main components of the lubrication system are the oil baths, the grease fittings, and the relief fittings. Some rotary table assemblies have oil bath lubrication only. Some use grease lubrication only, and some models use both an oil bath and grease for lubrication. Crew members should follow a regular lubrication schedule based on the manufacturer's guidelines. Always refer to these guidelines for more information on the lubricating system.

Oil Bath

An oil bath is the oil inside a compartment in the base of the rotary table assembly (figs. 15, 8). The oil surrounds the powerful and rapidly rotating table and drive shaft. The oil lubricates the equipment and cleans and cools it. A modern rotary has either one or two oil reservoirs built into it. Check the oil levels by using a dipstick. Also—

1. check the oil level according to a regular schedule,
2. maintain the correct level by adding oil when needed, and
3. keep the oil seals in good condition and replace them when they become old or damaged.

Occasionally, the oil in an oil bath can become contaminated. If it does—

1. correct the condition that caused contamination,
2. drain and properly discard the old oil in accordance with your company's disposal policies,
3. clean the reservoir of foreign matter, and
4. refill the reservoir with oil of the grade specified by the manufacturer.

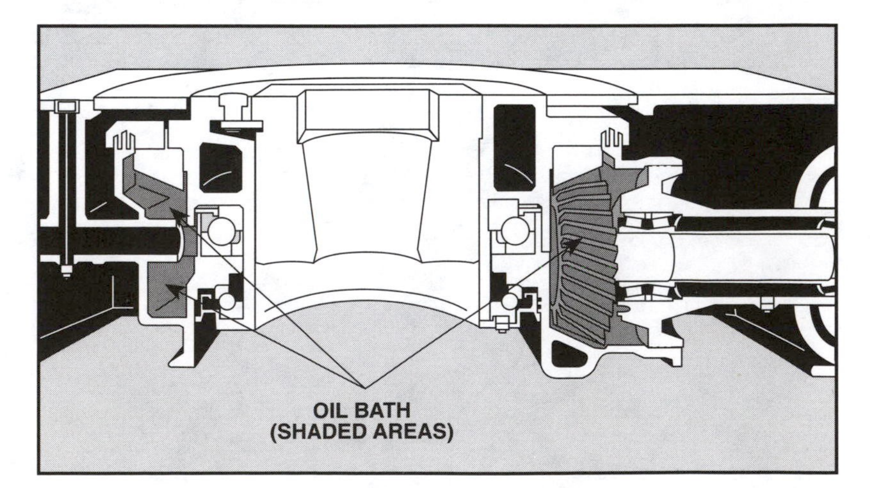

Figure 15. Oil bath lubrication

Grease Fittings

Fittings are openings in the base that allow righands to insert a grease gun and apply lubricant to moving parts inside the base (fig. 16). Manufacturers install the fittings near the parts needing lubrication, or at a convenient central point. A combination of oil bath and grease lubricates some rotaries. In these cases, a lubrication panel on the rotary holds grease fittings, the fill line for the oil reservoir, and the dipstick to check oil level (see fig. 13a).

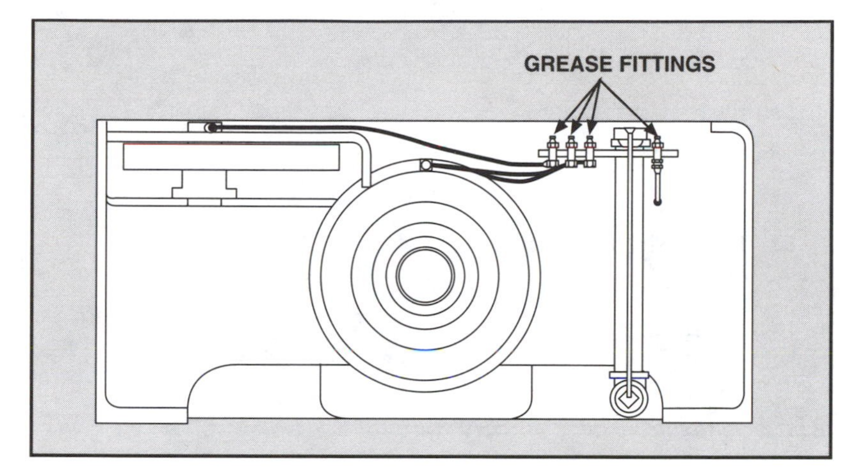

Figure 16. Grease fittings (top view of rotary table assembly)

Relief Fittings

Relief fittings are openings that unload excess lubricant. One place you may find a relief fitting is under the drive shaft of the rotary table assembly (fig. 17). Do not confuse the relief fittings on your rotary table with the grease fittings.

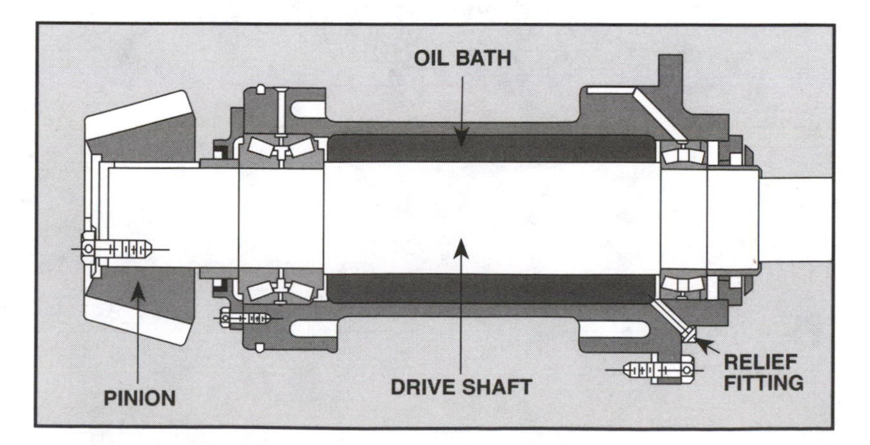

Figure 17. Relief fitting at base of rotary drive shaft

To summarize—

The rotary table assembly is a rotating machine housed inside a rectangular steel box. The assembly has an opening in the middle for the kelly and drill stem.

The rotary table assembly

- Rotates and transfers the turning motion to the kelly
- Suspends the weight of the drill string

The main parts of the rotary table assembly

- Base
- Turntable (or rotary table)
- Master bushing
- Drive-shaft assembly
- Drawworks sprockets
- Drive-shaft sprockets
- Locking devices

Maintenance procedures for the rotary table assembly

- Lubrication
- Cleaning
- Replacing mud seals

Installation of the rotary table assembly

- Check oil level
- Check grease fittings
- Position the rotary table under the traveling block
- Align the drive shaft and drawworks sprockets
- Check tension on the drive chain

▼
▼
▼

Master Bushing, Kelly Bushing, and Slips

▼
▼
▼

On a conventional rotary rig (one without a top drive), three important devices fit inside the rotary table: (1) the master bushing, (2) the kelly (or drive) bushing, and (3) the slips. The rig uses the master bushing and the kelly bushing during drilling. The rotary table assembly cannot turn the kelly directly. Instead, the master bushing and the kelly bushing transfer the rotary table's motion to the kelly.

Crew members use the slips when drilling stops. The slips hang (suspend) the drill stem in the rotary table assembly when righands make a connection, or when they trip the drill stem in and out of the hole.

Master Bushing

Definition

The master bushing is a rugged steel cylinder. It sits inside the turntable, which turns it (see fig. 5). The master bushing then turns the kelly bushing during normal drilling. The master bushing has a tapered surface for the slips. This surface is either part of the bushing itself, or it is a removable inner bowl that is separate from the master bushing (fig. 18). Manufacturers also provide the master bushing with a way to drive the kelly bushing. The two ways to drive the kelly bushing are the four-pin drive and the square drive.

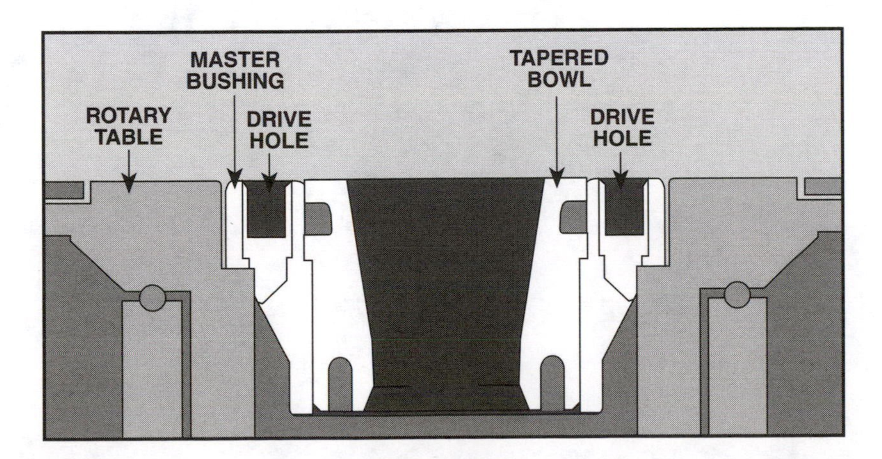

Figure 18. Tapered bowl that fits inside the master bushing

In four-pin-drive kelly bushings, four pins on the kelly bushing fit into corresponding holes in the master bushing (fig. 19). When the master bushing turns, the kelly bushing's pins also turn the kelly bushing. In square-drive kelly bushings, the kelly bushing has a square-shaped base. This square base fits into a corresponding square-shaped opening (receptacle) in the master bushing (fig. 20). When the master bushing turns, the kelly bushing also turns because its base is firmly inserted into the master bushing.

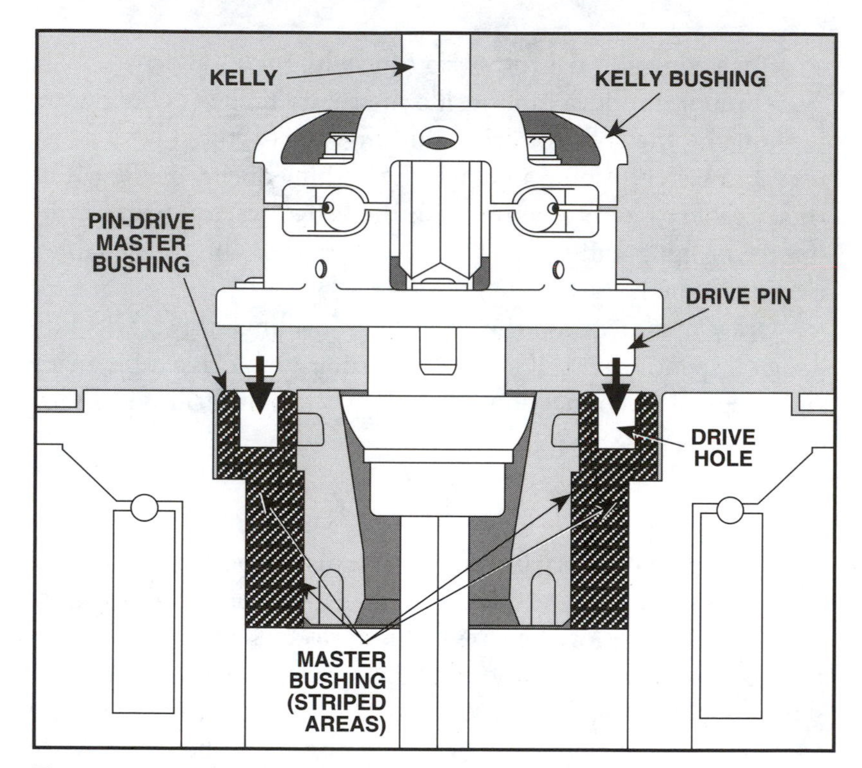

Figure 19. Four-pin-drive kelly bushing

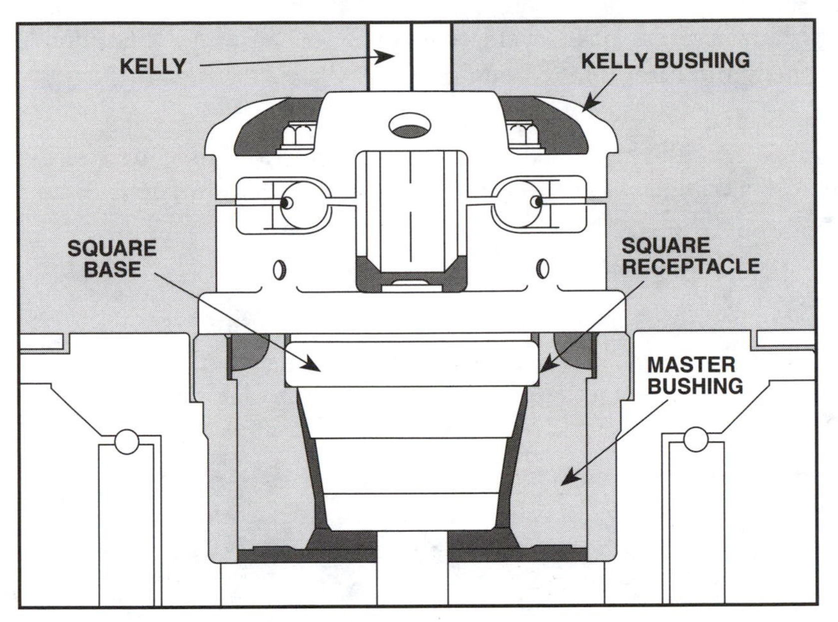

Figure 20. Square-drive kelly bushing

Functions

The master bushing performs two jobs:

1. During drilling, it connects the rotary table to the kelly bushing and transfers rotation from one to the other.
2. When drilling stops, the master bushing holds the slips (fig. 21). (Crew members remove the kelly bushing from the master bushing before setting the slips.)

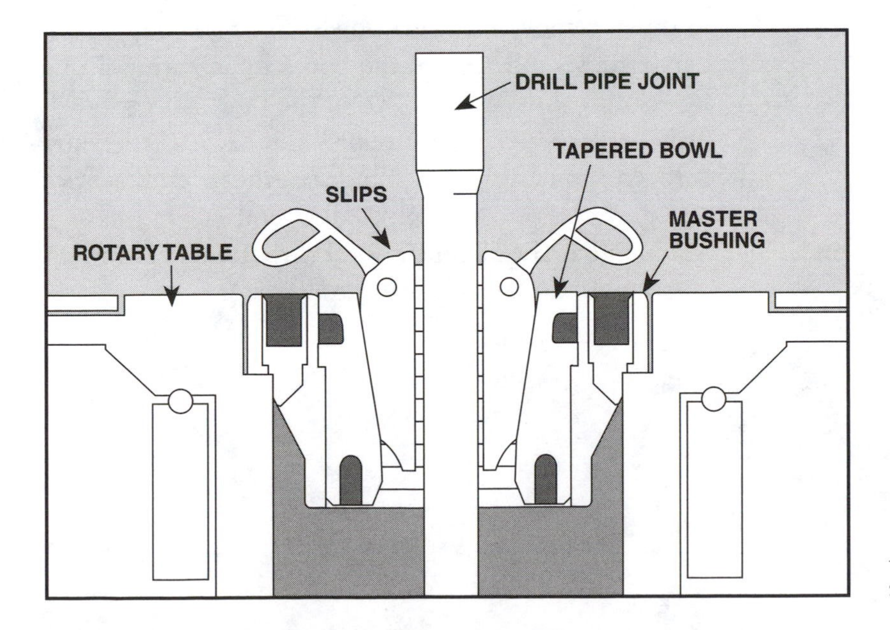

Figure 21. Slips in the master bushing hold the pipe.

Construction

Manufacturers make several types of master bushings. They divide them into categories based on—

1. construction,
2. the type of drive, or the connection, used to join the master bushing to the kelly drive bushing, and
3. size.

Design

Manufacturers make three kinds of master bushings:

1. *Split-construction type.* This type has two halves, each of which has a taper for the slips (fig. 22).

Figure 22. Split master bushing

2. *Solid-construction type.* Manufacturers make this bushing as a single piece. They machine the taper for the slips into two removable insert bowls (fig. 23). Crew members remove and replace the bowls by inserting lifting hooks into slots in each side of the bowl. Insert bowls provide the taper that supports the back of the slips. Also, crew members can change out the insert bowls for different sizes of drill pipe, drill collars, and casing. Table 1 gives the dimensions for three pipe sizes and the corresponding insert bowl sizes.

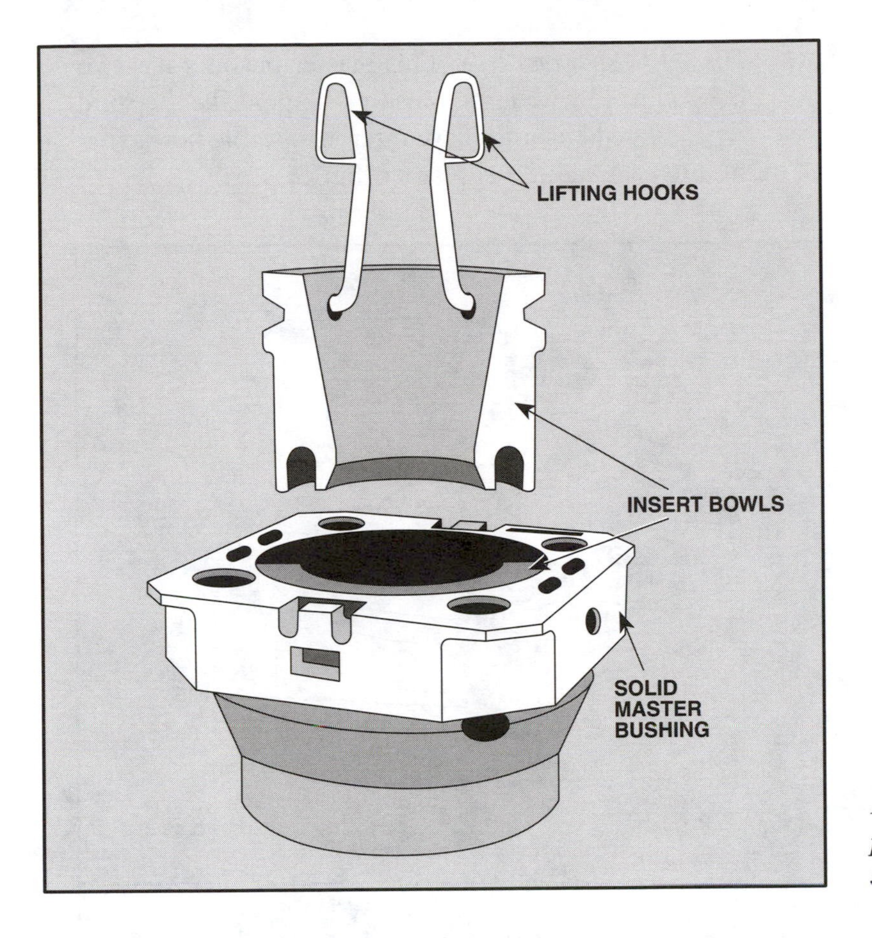

Figure 23. Solid, or single-piece, master bushing with split, or two-piece, insert bowl

Table 1
Insert Bowl Sizes

Pipe or Casing Size		Bowl	J		K	
Inches	Millimetres	No.	Inches	Millimetres	Inches	Millimetres
2⅜–8⅝	60–219	3	14⅜	365	10⅛	257
9⅝–10¾	244.5–273	2	16¼	413	12¼	311
11¾–13⅜	298.5–340	1	19	483	15	381

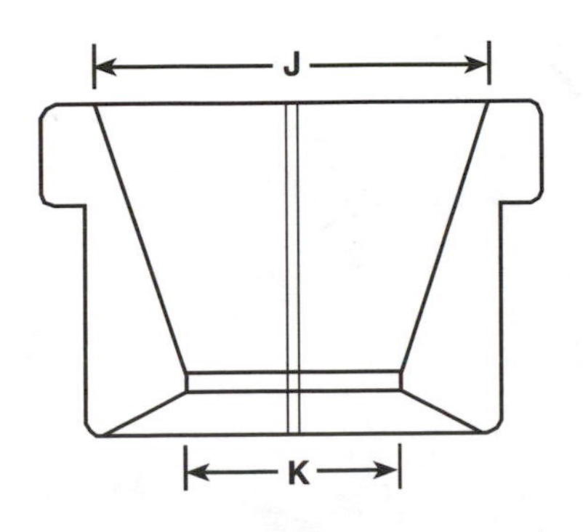

INSERT BOWL

3. *Hinged-construction type*. This master bushing has two halves linked with large pins in a hinge. Like the solid type, hinged bushings have interchangeable bowls that fit into each half for slip taper (fig. 24).

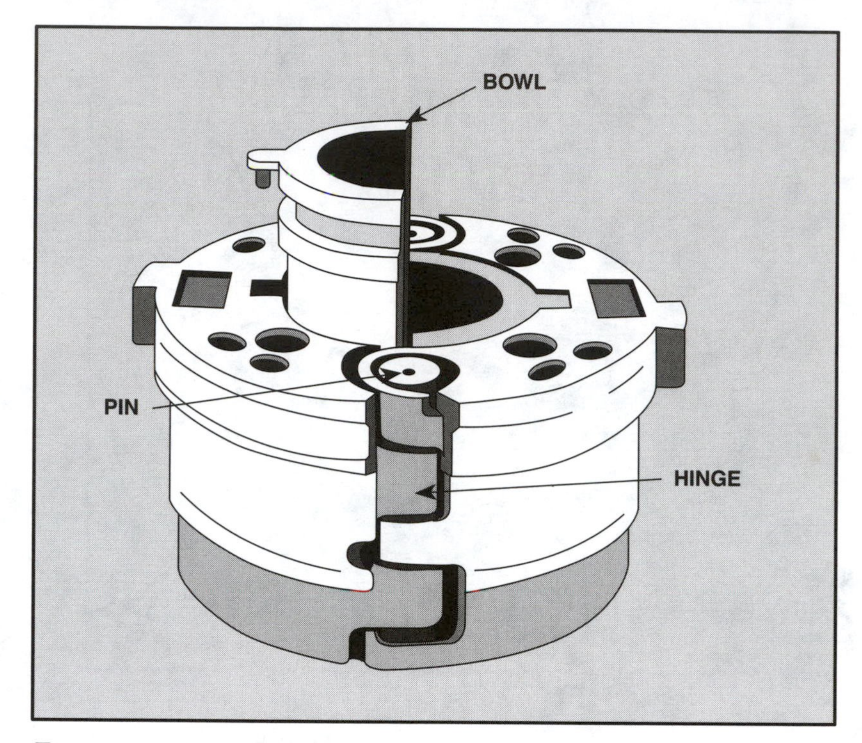

Figure 24. Hinged master bushing

Drives

A master bushing must rotate the kelly bushing so that the kelly bushing can rotate the kelly. As stated earlier, master bushings drive kelly bushings in one of two ways. Either—

1. four pins on the kelly bushing fit into four holes in the master bushing (see fig. 19), or
2. a square shape on the bottom of the kelly bushing mates with a square opening in the master bushing (see fig. 20).

Square-drive master bushings are lighter and less expensive than pin-drive bushings. Square-drive bushings cannot, however, support the weight of a very long and heavy drill stem. Nor can they accommodate the long slips that are necessary for a long string of pipe. So, land rigs that drill mostly shallow wells are more likely to have square drives. Square drives have a square opening that measures 13⁹⁄₁₆ inches (344.49 millimetres) on each side. The square opening is 4 inches (101.6 millimetres) deep to accept the kelly bushing's square bottom. The length of its taper is 8¹³⁄₁₆ inches (224 millimetres).

Pin-drive master bushings allow the rotary table to support more weight and can accommodate long slips. Contractors therefore use them on rigs that are drilling deep wells. Pin-drive bushings extend the length of the slip taper to 12¾ inches (324 millimetres). The pin drive's longer taper allows crew members to use long slips that support heavy weights (fig. 25).

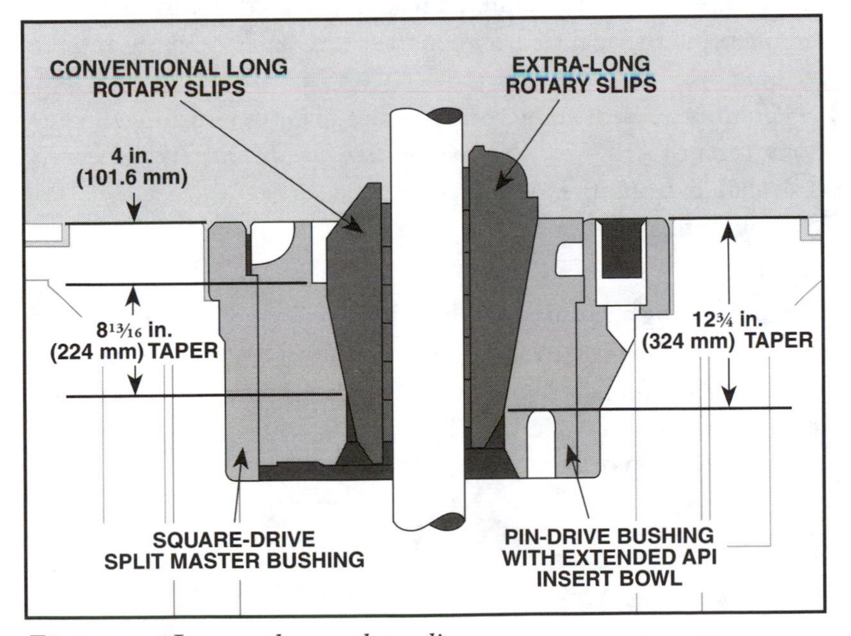

Figure 25. Long and extra-long slips

Size

API standards specify the dimensions of rotary tables and master bushings. Manufacturers make five rotary tables with openings whose internal diameters (IDs) meet API standards. These rotary table IDs are 17½, 20½, 27½, 37½, and 49½ inches (444.5, 520.7, 698.5, 952.5, and 1,257.3 millimetres). Since the master bushing fits inside the rotary table, the ID of the rotary table opening determines the outside diameter (OD) of the master bushing. For example, a rotary table with a 17½-inch (444.5-millimetre) ID requires a master bushing with a 17⁷⁄₁₆-inch (442.9-millimetre) OD. Usually, however, rig workers and owners use the rotary table's ID to designate the master bushing needed for that particular table. For example, if the rig's rotary table ID is 17½ inches (444.5 millimetres), the rig owner orders a 17½-inch (444.5-millimetre) master bushing, even though the master bushing's actual OD is 17⁷⁄₁₆ inches (442.9 millimetres).

Manufacturers make square-drive and pin-drive master bushings to fit four of the five rotary table sizes. A single-piece 37½-inch (952.5-millimetre) or 49½-inch (1,257.3-millimetre) master bushing would be so heavy that a rig's air hoist could not move it. Manufacturers therefore make the 37½-inch (952.5-millimetre) bushing in two hinged pieces. The rig's air hoist can safely move the bushing one half at a time. If a rig has a 49½-inch (1,257.3-millimetre) rotary table, the owner buys the 37½-inch (952.5-millimetre) hinged master bushing and an adapter. To install the master bushing, crew members put the adapter between the rotary table opening and the bushing.

For more details about rotary table and master bushing dimensions, refer to API Spec 7K, *Specification for Drilling Equipment*. It is available from the API Order Desk, 1220 L Street, NW, Washington, DC 20005.

Additional Equipment

Master bushing accessories include lifting slings for removing and installing the master bushing and the bowls (fig. 26), and a bit breaker adapter (fig. 27). The adapter makes it possible for the crew to install the bit breaker. The bit breaker holds the bit while the crew makes it up or breaks it out.

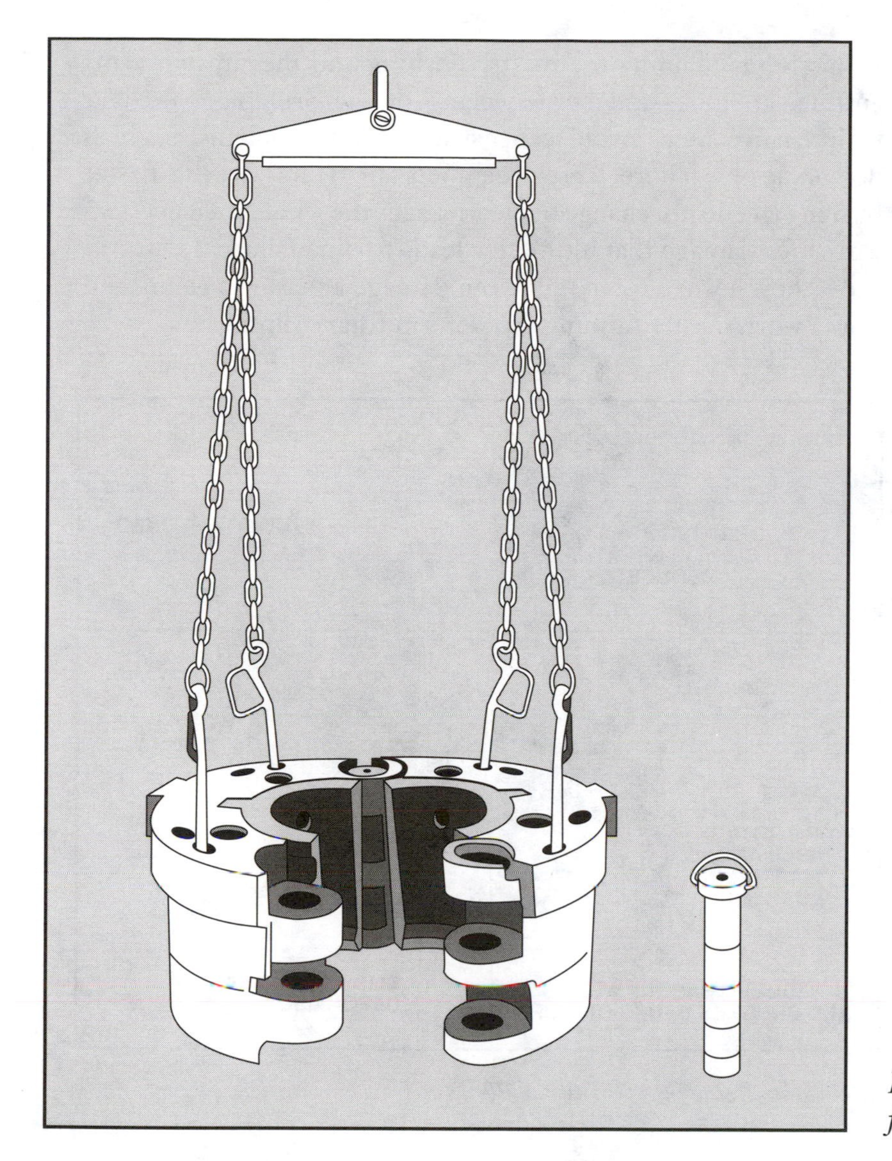

Figure 26. Lifting slings for a master bushing

Figure 27. Bit breaker adapter for a pin-drive master bushing

Maintenance

Properly maintaining the master bushing and the slips prevents cutting, gouging, and bottlenecking of the drill pipe. Regular maintenance helps avoid expensive drilling problems, such as downhole pipe failure. If crew members fail to lubricate the master bushing and do not change out worn parts, the whole rotary system can suffer damage that ultimately leads to pipe failure. Figure 28 shows how a worn rotary table can damage a master bushing and how a worn master bushing can deform rotary slips.

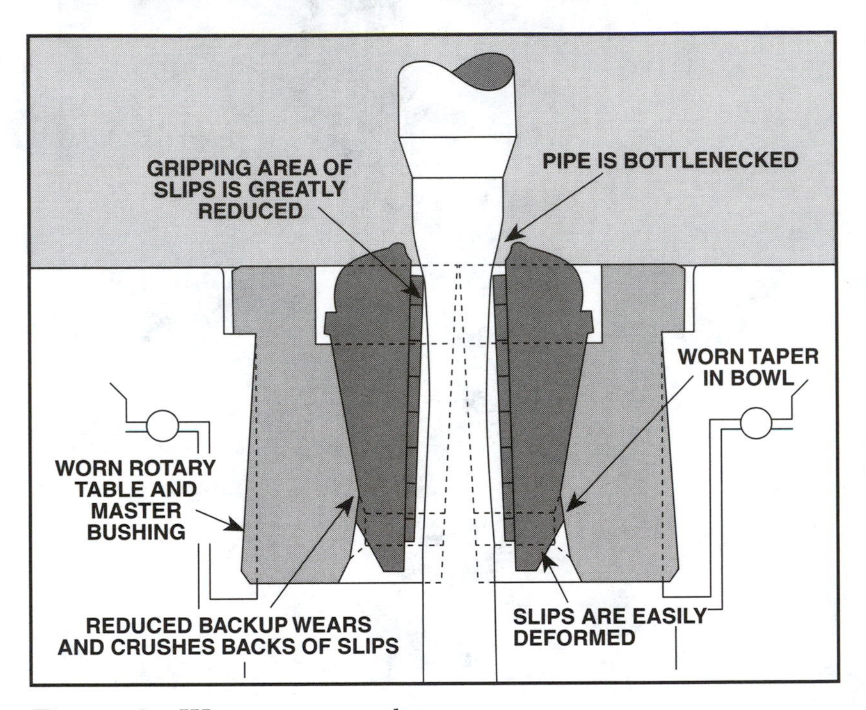

Figure 28. Worn parts cause damage.

Figure 29 illustrates another example of how one piece of worn equipment affects another. In the illustration, crew members installed a new split master bushing in a rotary table with too much ID wear. The bad fit caused the bushing to spread at the bottom. To solve the problem, the crew can either build up the ID of the rotary table or use a solid master bushing. A solid bushing does not depend on the table's ID for support.

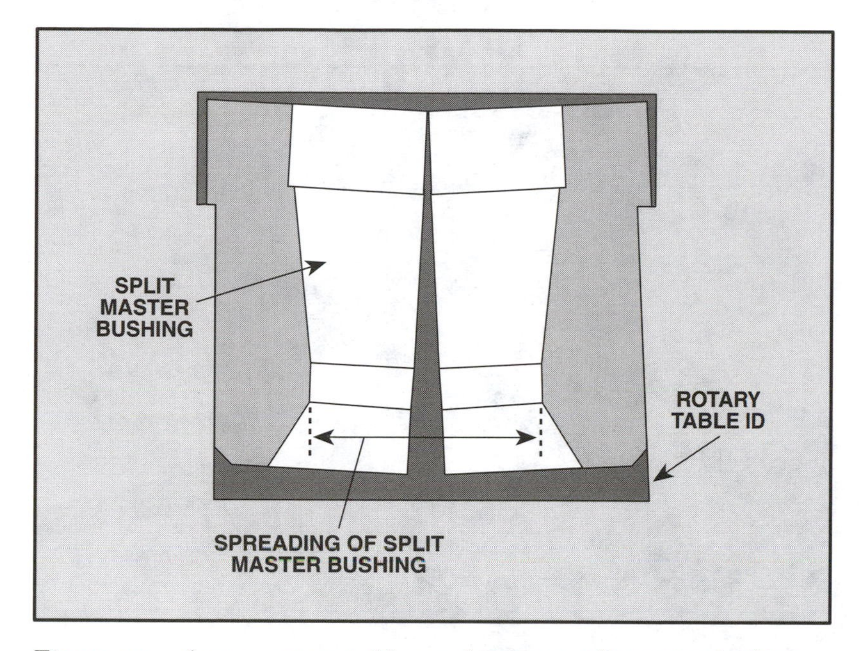

Figure 29. A worn rotary table can damage a split master bushing.

Kelly Bushing
Definition

The kelly bushing (also called the drive bushing) turns the kelly. It fits inside the rotary table's master bushing (fig. 30). The kelly bushing's dome-shaped metal housing has an opening in the middle for the kelly, four pins or a single square drive on the bottom, and several roller assemblies inside. Roller assemblies are removable units that allow the kelly to move up and down inside the kelly bushing. These assemblies contain bearings, seals, roller pins, and rollers.

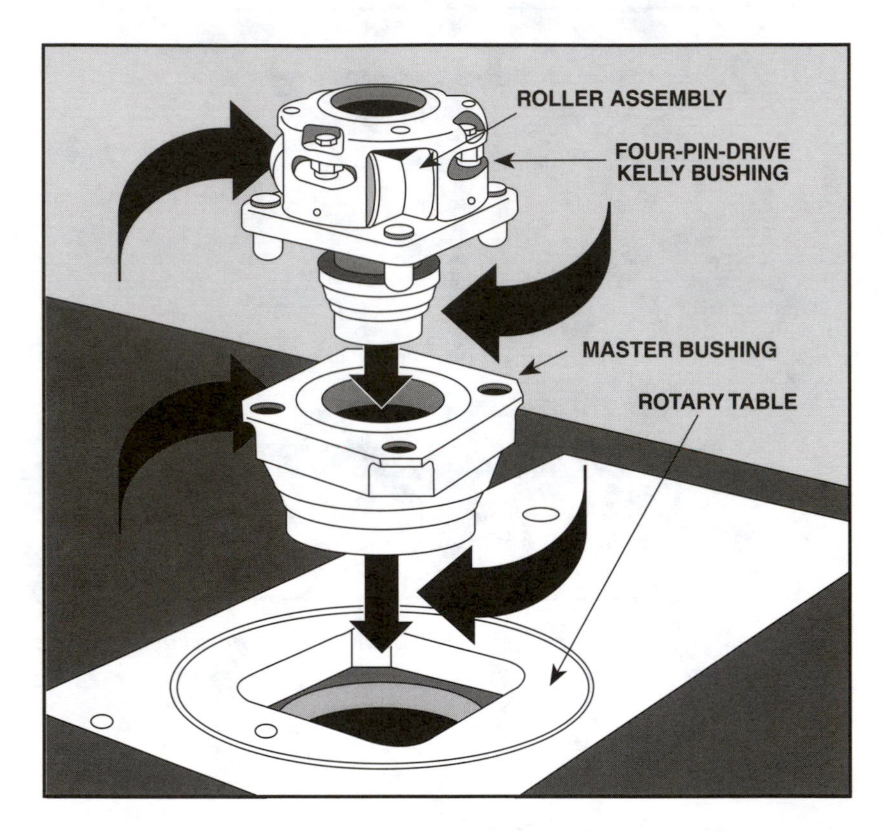

Figure 30. Kelly bushing, master bushing, and rotary table

Functions

The kelly bushing (fig. 31)—

1. transfers turning motion from the rotary table's master bushing to the kelly, and
2. allows the kelly to move up and down freely. The kelly moves down with the drill stem as the hole deepens. The driller lifts the kelly to allow the rotary helpers to add a new joint of drill pipe to the drill stem.

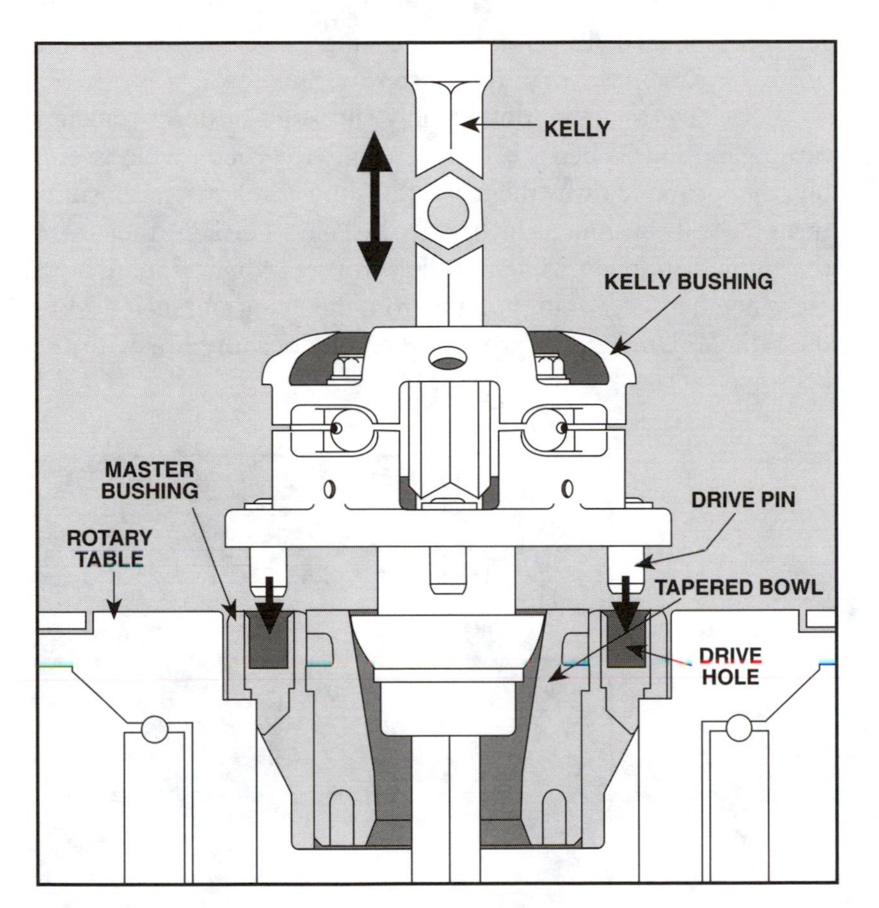

Figure 31. Kelly bushing and master bushing

How the Kelly Bushing Works

The kelly bushing transfers turning motion to the kelly by engaging the master bushing. Since manufacturers make two types of master bushing drives (the four-pin and the square), they also make two types of kelly bushing drives to fit them:

1. The manufacturer puts four drive pins on the bottom of the kelly bushing. Each pin fits into holes bored into each corner of the master bushing.
2. The manufacturer attaches a square piece on the bottom of the kelly bushing. The square bottom mates with a square recess in the master bushing.

Let's take a closer look at how the kelly bushing transfers turning motion. When the driller engages the rotary table assembly, the turntable turns the master bushing. The master bushing turns the kelly bushing, which turns the kelly. The kelly then turns the entire drill string. When drilling stops, as when tripping pipe, crew members break out the kelly from the string and set it back in the rathole. During this process, the kelly bushing lifts with the kelly and sets back with it (fig. 32).

Figure 32. Kelly and kelly bushing set back in the rathole

The kelly bushing also lets the kelly move freely up or down when the rotary is turning or when it is stationary. Inside the housing, specially designed rollers fit the flat sides of the kelly (fig. 33).

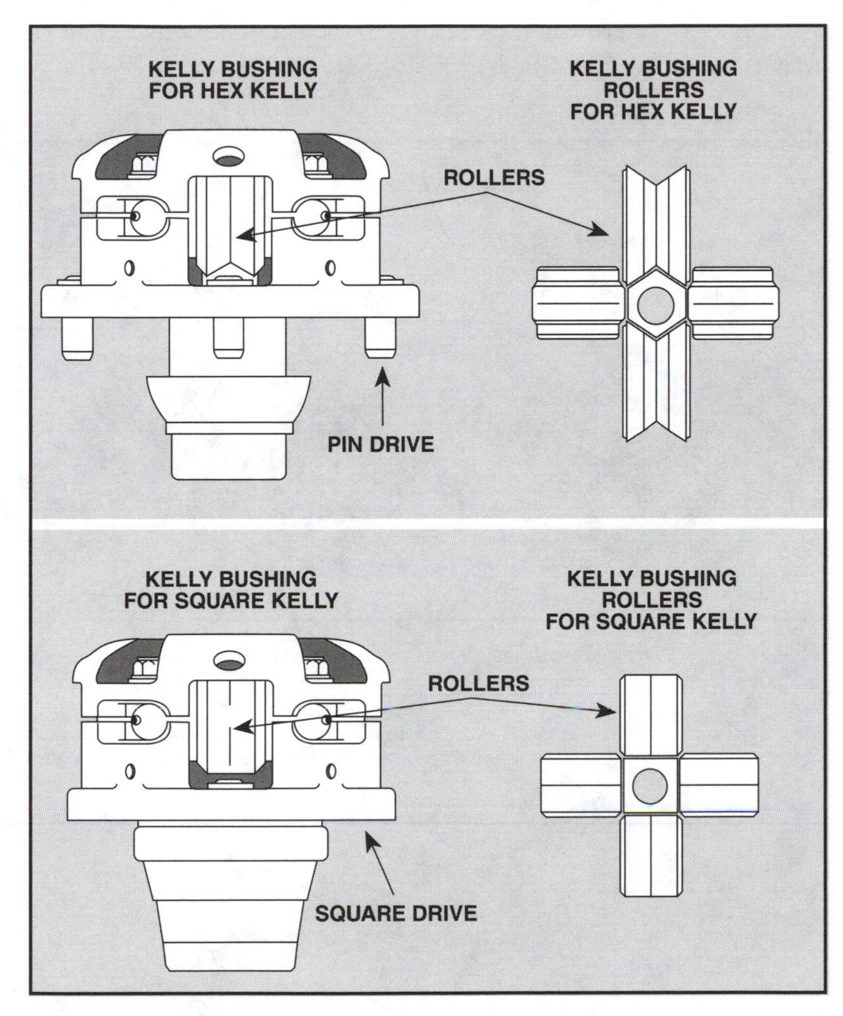

Figure 33. Rollers fit the kelly's shape.

These rollers allow vertical movement. They are housed in units called roller assemblies, which contain roller pins (fig. 34). Roller bearings, lubricant, and lubricant seals complete the assembly. The harsh demands of drilling put wear on the rollers. The assemblies make it easy for crew members to remove the rollers for repair or to accommodate the different kelly shapes.

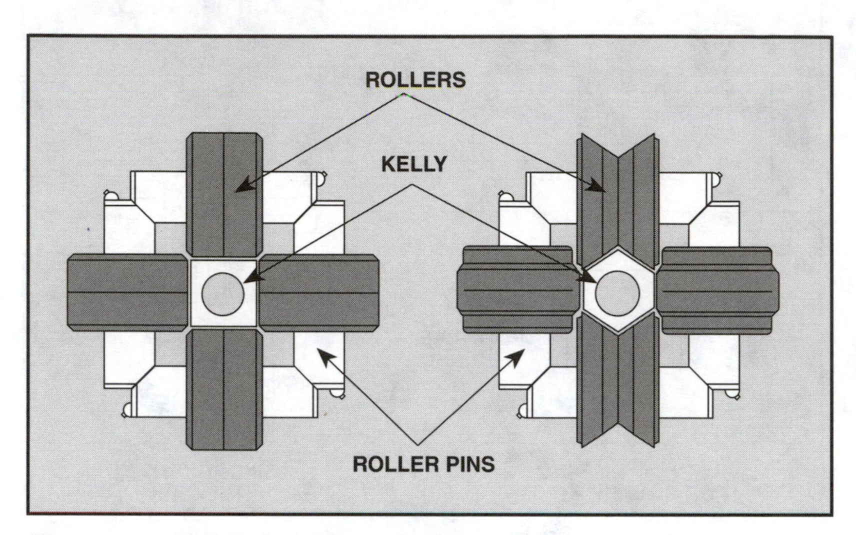

Figure 34. Roller assemblies

Design

Manufacturers design kelly bushings for two shapes of kelly and with one or two rows of rollers. Kellys come in two shapes: square and hexagonal. If you cut a kelly across the middle to take a cross section of it, it would be either square (have four sides) or hexagonal (have six sides) (fig. 35). Since kelly bushings fit closely around the kelly, manufacturers make the kelly opening (the place where the kelly goes through the bushing) either square or hexagonal. At least one type of kelly bushing is available that crew members can adapt to fit both shapes of kelly. They simply change the rollers in the drive to mate with a square or a hexagonal kelly.

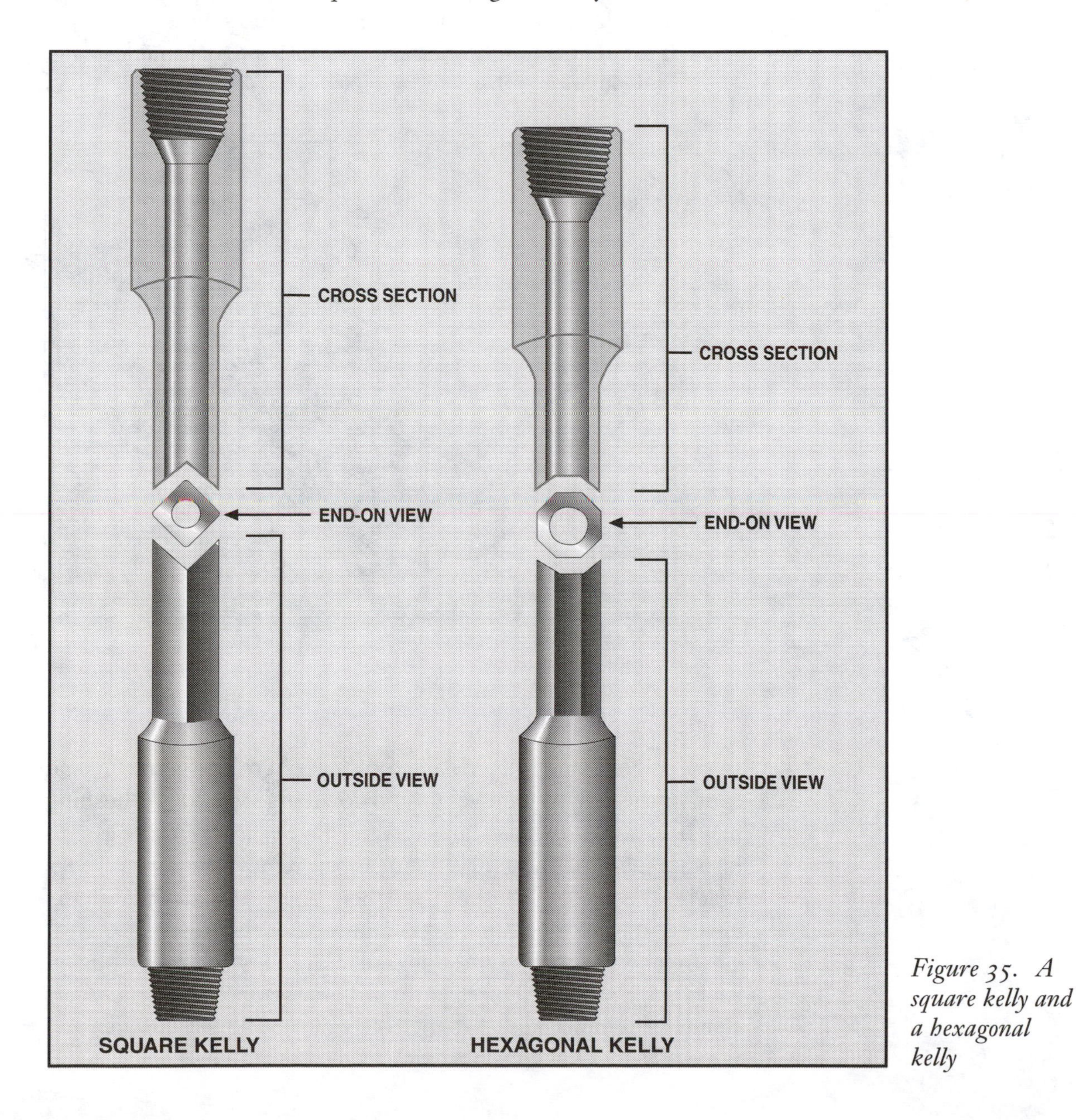

Figure 35. A square kelly and a hexagonal kelly

Some kelly bushings have a lock assembly that secures it to the master bushing (fig. 36). This lock is useful on floating offshore rigs if the rig is not using a top-drive system. In such cases, crew members lock the kelly bushing to the master bushing. When locked, the kelly bushing cannot come out of the master bushing as the floating rig moves up and down with the waves.

Manufacturers mount the rollers in roller assemblies (see fig. 34). Roller assemblies contain rollers that guide the kelly and roller bearings that turn inside the rollers. A roller pin, which is a shaft, rotates inside the bearings. They may arrange the assemblies on a single horizontal plane, or on double planes. In a single-plane type, the kelly moves through one layer of rollers. In a double-plane type, the kelly moves through two levels of rollers.

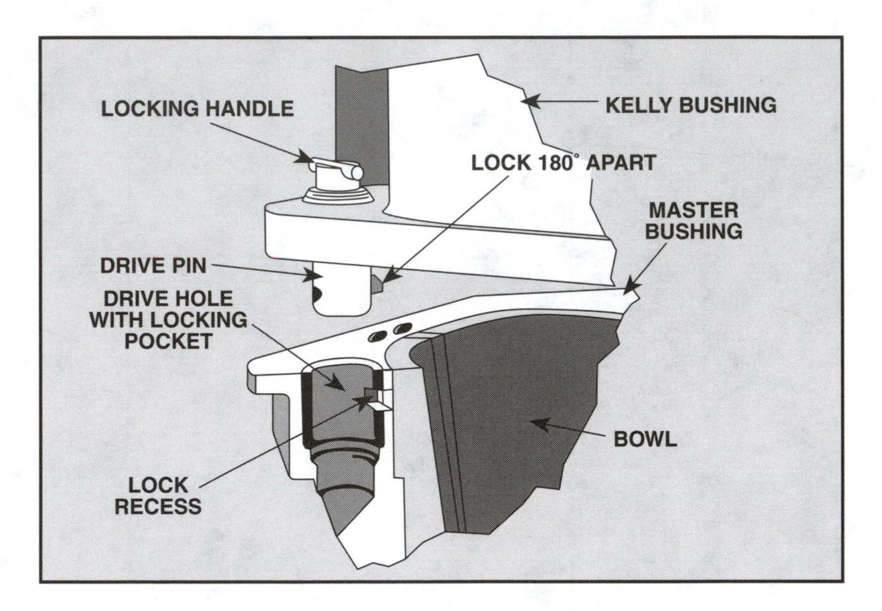

Figure 36. Lock assembly for kelly bushing and master bushing

Single-Plane Roller Design

In the single-plane roller design, one layer of rollers surround and support the kelly as it moves up and down (fig. 37). A kelly bushing may have a split- or solid-body casting. To replace the roller assemblies in a split-body bushing, crew members remove the nuts and bolts holding the top of the bushing, and then remove the cover. With the cover removed, they can inspect and replace the rollers.

In a solid-cast body, the manufacturer fits the roller pins in replaceable sleeves. To reach the roller assembly, crew members remove retaining pins, take the roller pins out from the side, and remove the rollers from the back.

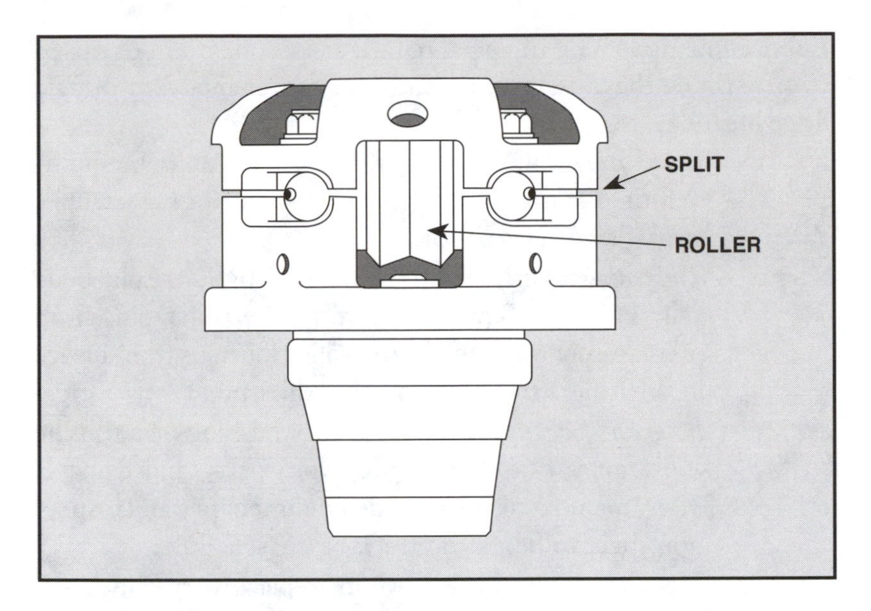

Figure 37. Single-plane rollers in a split-body bushing

Double-Plane Roller Design

Double-plane roller kelly bushings use two rollers for each driving surface on the kelly (fig. 38). The manufacturer puts the rollers, the roller pins, and the bearings in a cage. Each cage is then inserted into the kelly bushing. (Typically, four cages are used.) When required, crew members can remove the cages from the bushing body. The two-roller-plane bushing for hexagonal kellys is adjustable for kelly and bushing wear. The two-roller bushing for square kellys is not adjustable.

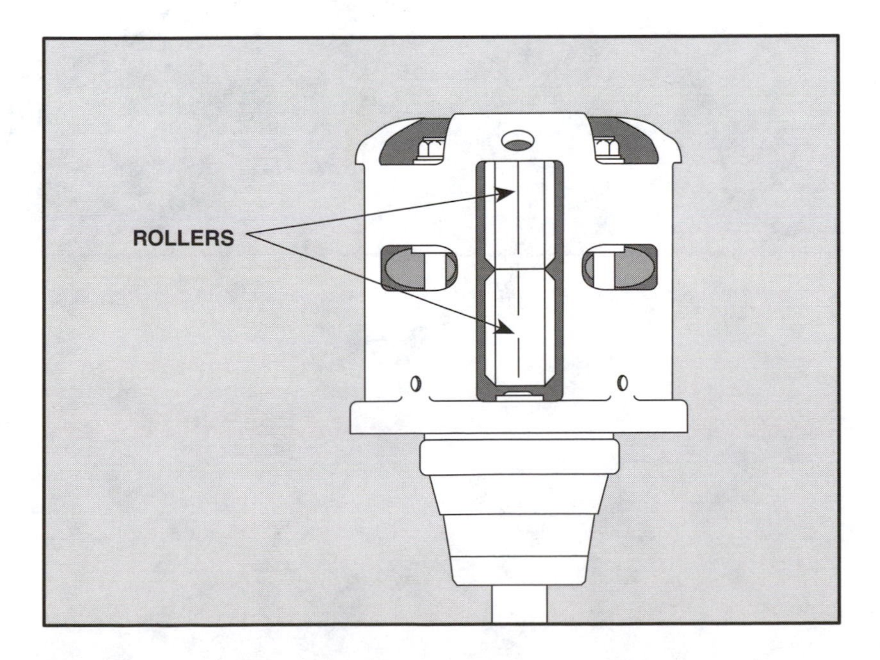

Figure 38. Double-plane rollers

Maintenance

Lubrication, tightening top nuts, regular inspections, and replacing worn parts are the most crucial points of maintenance for the kelly drive bushing.

1. *Lubrication.* Apply grease with a grease gun to the built-in fittings in the bushing. Grease regularly—for example, every day or every tour.
2. *Tightening top nuts.* Keep the top nuts tight on split-body kelly bushings. This action keeps the roller pins from prematurely wearing out the roller journals (the surface in the rollers that turns on the roller pins).
3. *Inspecting for wear.* Inspect the kelly bushing on a routine and regular basis for fatigue, worn parts, and damage. Many manufacturers provide a maintenance and inspection sheet to help keep track of the schedule.
4. *Replacing worn parts.* Remove and replace worn kelly bushing parts under the supervision of the driller (figs. 39 and 40).

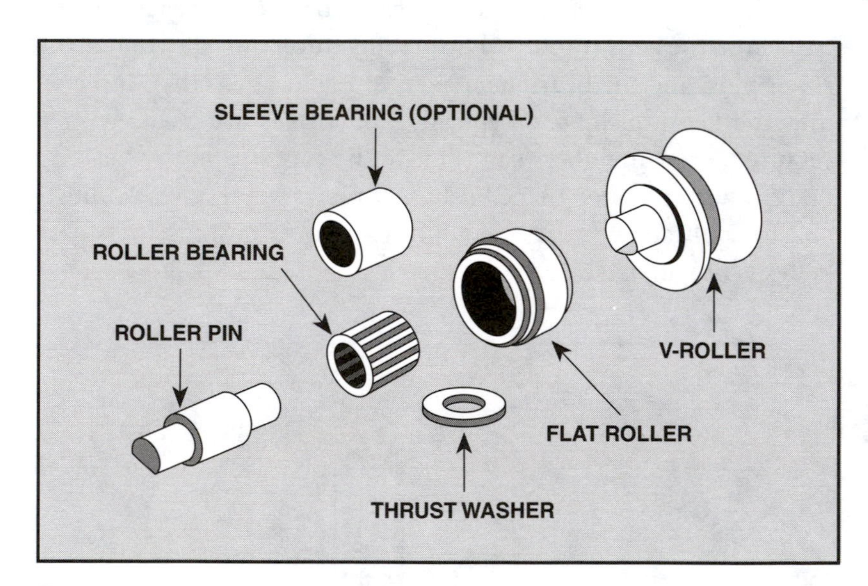

Figure 39. Kelly bushing roller assembly

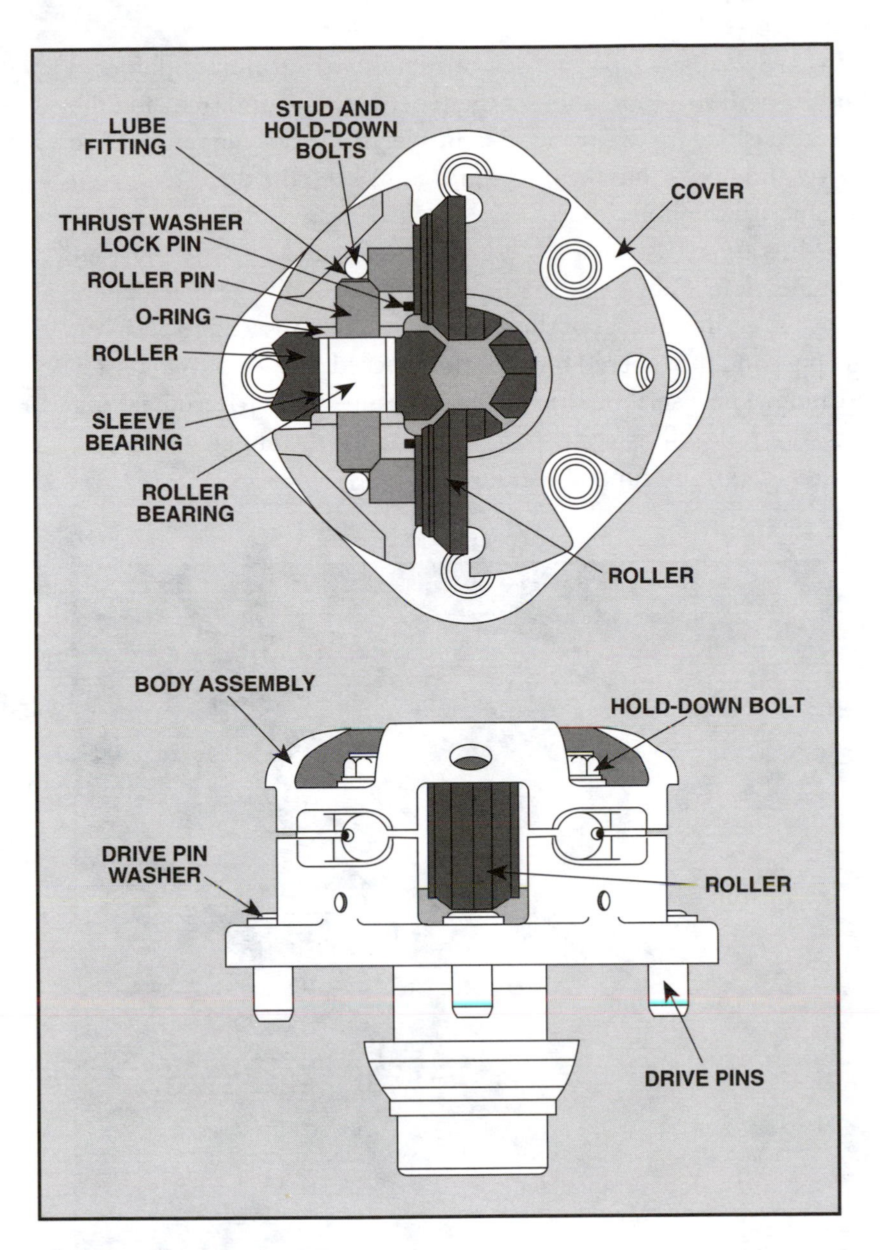

Figure 40. Kelly bushing components

Slips

Definition

Slips are wedge-shaped pieces of metal with gripping elements (inserts or dies). Floorhands set them between the drill stem and the master bushing (fig. 41). With the slips set, the hands can remove the swivel and hook from the drill stem. Slips keep the drill stem from falling into the hole.

Slips are necessary when crew members make up or break out a connection. They place the slips around drill pipe or drill collars (they use a different set of slips for each). The slips wedge between the pipe and the tapered bowl or the tapered surface of the master bushing. The inserts on the slips grip the pipe to keep it from falling into the hole. Slips consist of three or more hinged segments, inserts (dies), and hinged handles.

Figure 41. Crew members setting slips

Function

When crew members set the slips around the drill stem in the master bushing, the slips grip the pipe without damaging it. With the slips set, the rotary table assembly suspends the entire drill stem.

How Slips Work

When normal drilling is in progress on a conventional rig, the blocks, hook, and swivel support the kelly and the drill stem attached to it. Floorhands often disconnect the swivel and kelly from the drill stem to add or remove pipe from the drill stem. To disconnect the swivel and kelly from the drill stem, crew members use slips. The slips take over for the hook, swivel, and kelly by suspending the drill stem in the hole.

To use the slips, the driller first raises the drill stem until the kelly and the kelly bushing are above the rotary table assembly. The kelly pulls the top joint of drill pipe up through the rotary assembly and positions it inside the master bushing (fig. 42).

Figure 42. Top joint of drill pipe positioned inside the master bushing

The crew places the slips around the drill pipe (fig. 43), and the driller slowly lowers the pipe. When the pipe stops moving, the rotary helpers set the slips. The driller then eases off the drawworks brake (slacks off). This action puts all the drill stem's weight on the slips. The weight of the drill stem tightly wedges the slips into the master bushing (fig. 44). The inserts on the slips grip the pipe and keep it from falling into the hole. The slips and the rotary table assembly now suspend the drill stem. At this point, the crew can break out the kelly and the swivel and add another joint of pipe (fig. 45) or perform some other operation.

Figure 43. Slips around drill pipe

Figure 44. Slips set in master bushing

Figure 45. Crew members breaking out pipe

The slips hold the drill stem securely because they use the weight of the drill stem itself to transfer the downward (axial) force of the hanging drill string to a sideways (transverse) force. This transverse force wedges the drill stem in the slips against the master bushing (fig. 46).

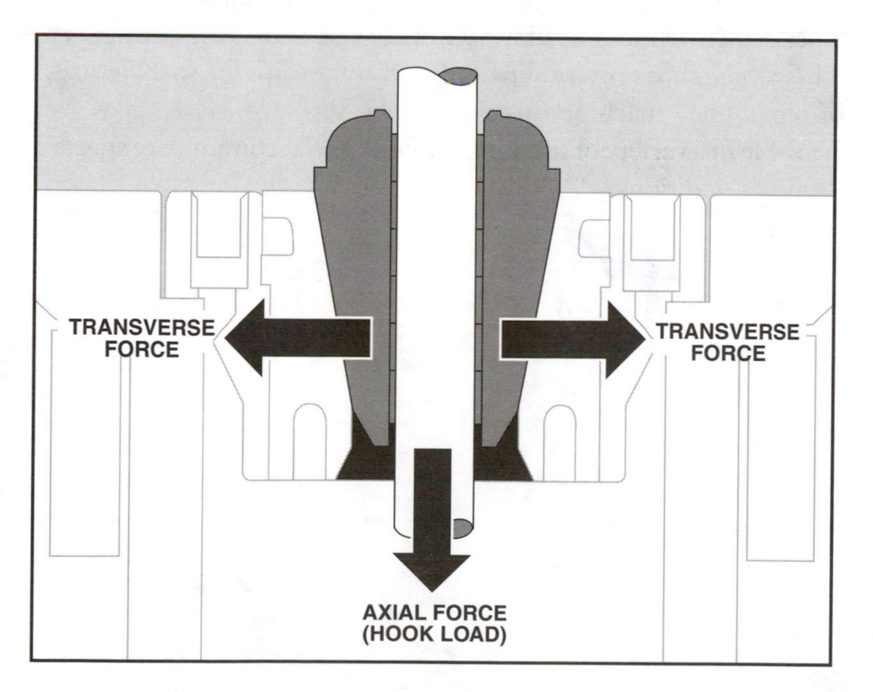

Figure 46. Slips transfer downward (axial) force to sideways (transverse) force.

Design

There are three kinds of slips—

1. rotary slips, which crew members use with drill pipe only (fig. 47);
2. drill collar slips, which crew members use with drill collars only (fig. 48); and
3. casing slips, which crew members use when running casing.

Drill collars are much larger in diameter than drill pipe. Drill collar slips therefore have many hinged segments to properly fit around the drill collar. Often, drill collar slips use special button-shaped dies as gripping elements. The manufacturer mounts the buttons on the insert that goes into each slip segment.

Casing is also larger in diameter than drill pipe. Casing slips are therefore specially made to suspend only casing in the rotary table.

Slips have several hinged body segments. Typically, rotary slips (see fig. 47) have three segments and drill collar slips (see fig. 48) have ten or more segments. The segmented design gives the slips flexibility without sacrificing strength.

The removable inserts slide inside slots in each slip segment. A special pin holds them there. A cotter key locks the pins in place. This pin-and-cotter arrangement ensures that the inserts stay in the slots, but also allows crew members to replace the inserts easily. Inserts wear as they suspend joint after joint of pipe, and are made to be discarded when worn out.

Manufacturers make the outside of the slips smooth, and they taper them to fit the tapered surface of the master bushing (see fig. 21). They design both rotary slips and drill collar slips for specific sizes of pipe. They machine this size on the slip face. Both types are available in a variety of sizes and capacities to accommodate special drilling conditions.

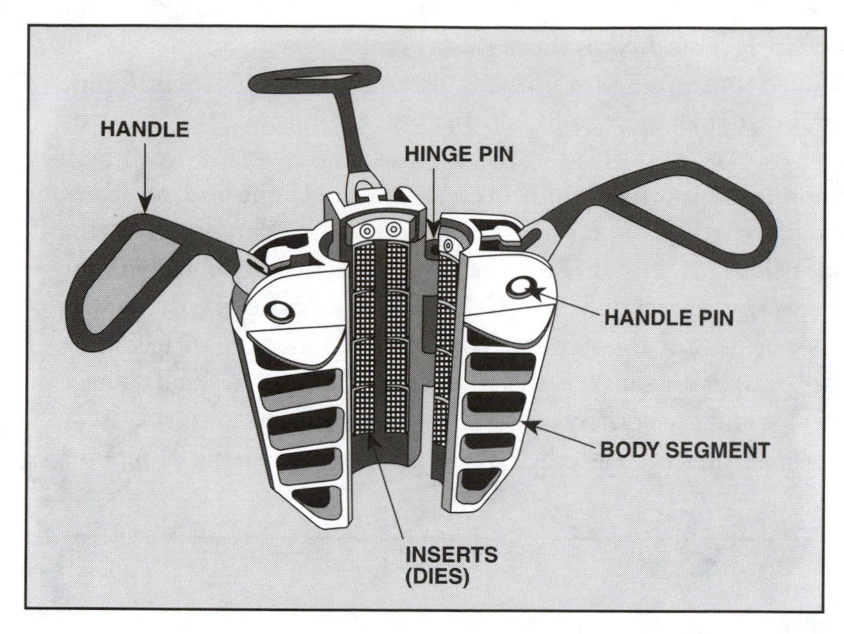

Figure 47. Rotary slips for use with drill pipe

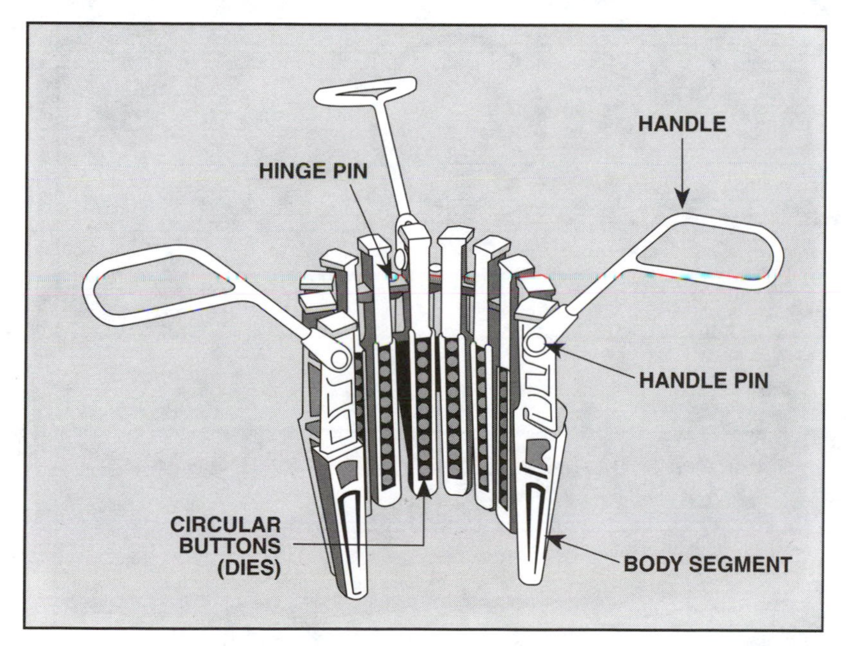

Figure 48. Drill collar slips

Size

Slip size is critical. Slips should always match the size of pipe being handled (fig. 49). For example, 5-inch (127-millimetre) drill pipe requires 5-inch (127-millimetre) slips. Toolpushers or company supervisors normally plan for the slip sizes a rig needs and order them in advance. Floorhands, however, should check the size machined onto the slip faces and inserts to make sure they are correct for the pipe being run. Slips that are too large do not make contact with the pipe all the way around (fig. 50). Poor contact can cause the pipe to drop and destroy the center part of the slips' gripping surface. The sides (ribs) of the slips can also crack. Slips that are too small damage the pipe and the corners of the slips. The slip ribs can crack. Also, too-small slips increase the risk of dropping the string of pipe.

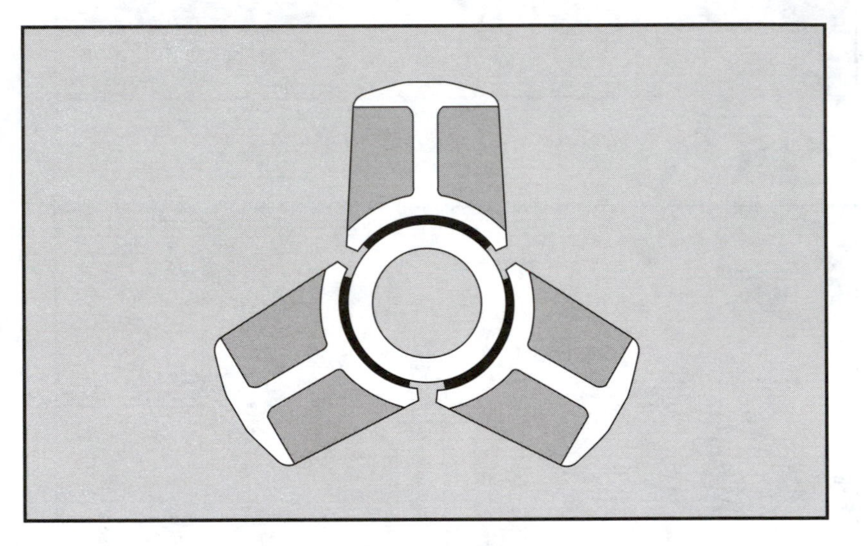

Figure 49. Properly sized slips uniformly grip the pipe.

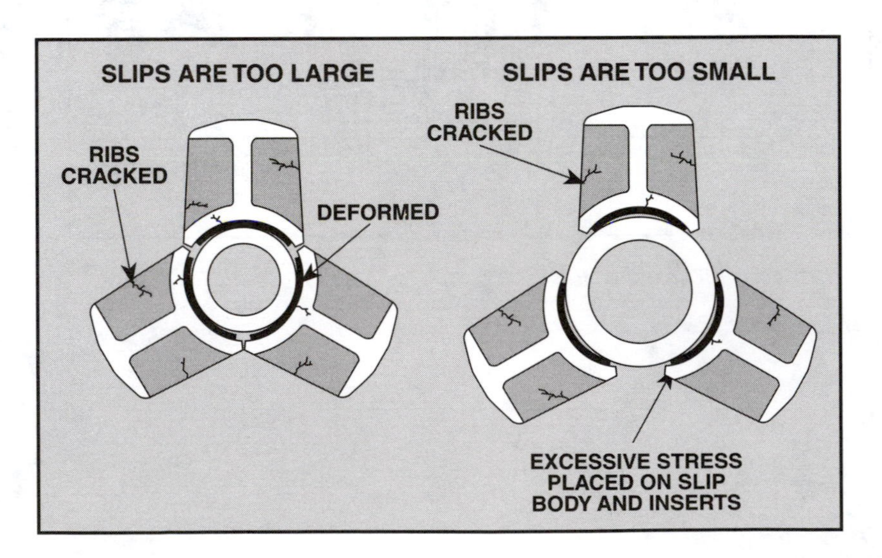

Figure 50. Slips that are the wrong size for the pipe being run may damage both the slips and the pipe.

Handling

Remember the following when setting and removing slips.

1. Never use the slips to stop pipe when the driller is lowering it. Let the driller stop the pipe with the drawworks brakes first. Then set the slips. Letting the slips ride the pipe down into the master bushing to stop them is an easy habit to fall into when tripping pipe. It is, however, an expensive habit that can result in permanent pipe deformation (fig. 51).
2. Never allow the slips to ride on the pipe when the driller raises the pipe out of the hole. This practice increases wear on the slips and could damage the pipe. It also risks injury if the moving pipe suddenly pulls the slips out of the rotary table. Instead, use the hinged handles to lift the slips before the driller raises the joint.

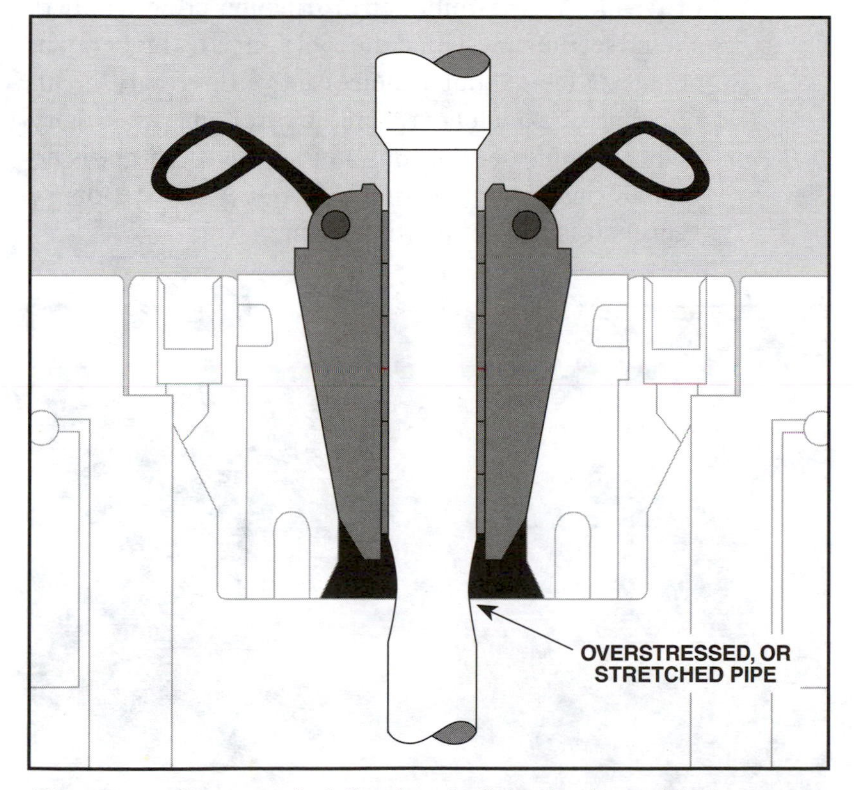

Figure 51. Pipe stretched (bottlenecked) by the slips suddenly stopping the drill stem

3. Set the drill stem in the slips so that the tool joint is as close to the rotary table as possible (fig. 52). If the floorhands set the pipe with the tool joint too far above the rotary table, the pull on the tongs needed to break out or make up the pipe could bend it. Imagine a newly planted tree. If you grab the little tree at its very top, you can easily bend it. On the other hand, if you grab the tree at the bottom of its trunk, near the point where the ground anchors its roots, you cannot bend it. API RP 7G, *Recommended Practice for Drill Stem Design and Operating Limits*, gives equations for determining the maximum height the tool joint can be above the slips to prevent bending. Refer to this booklet for more details. In short, the smaller in diameter the pipe is, the closer the tool joint should be to the rotary table. For example, with one type of 4½-inch (114.3-millimetre) drill pipe, crew members should set the slips so that the tool joint is no higher than about 3½ feet (about 1 metre) above the rotary. With one type of 4-inch (101.6-millimetre) pipe, the rotary helpers should set the slips so that the tool joint is no higher than about 2½ feet (about ¾ metre or 75 centimetres) above the rotary table.

Figure 52. Drill stem set in slips at the correct height above the rotary table

Repair

Worn slips are dangerous. Each crew member should visually check the slips and the inserts for wear and replace or repair potentially hazardous equipment. Worn inserts can cause the pipe to slip and bounce, which can kick the slips out of the master bushing. Worn inserts and damaged slips not only create a dangerous condition, but also increase the risk of losing the drill string down the hole.

When inserts become worn, replace all of them at once. Replacing only a few of the inserts can lead to trouble because old inserts have worn surfaces. As a result, the old inserts do not contact the pipe and carry little if any of the load. The new inserts thus dig into the pipe and damage it. Also, the load on the new inserts forces them into the body of the slips and damages them, too (fig. 53).

Handles, pins, and the body of the slips also take a lot of drill stem weight and shock. Regularly inspect these parts and replace them immediately if necessary.

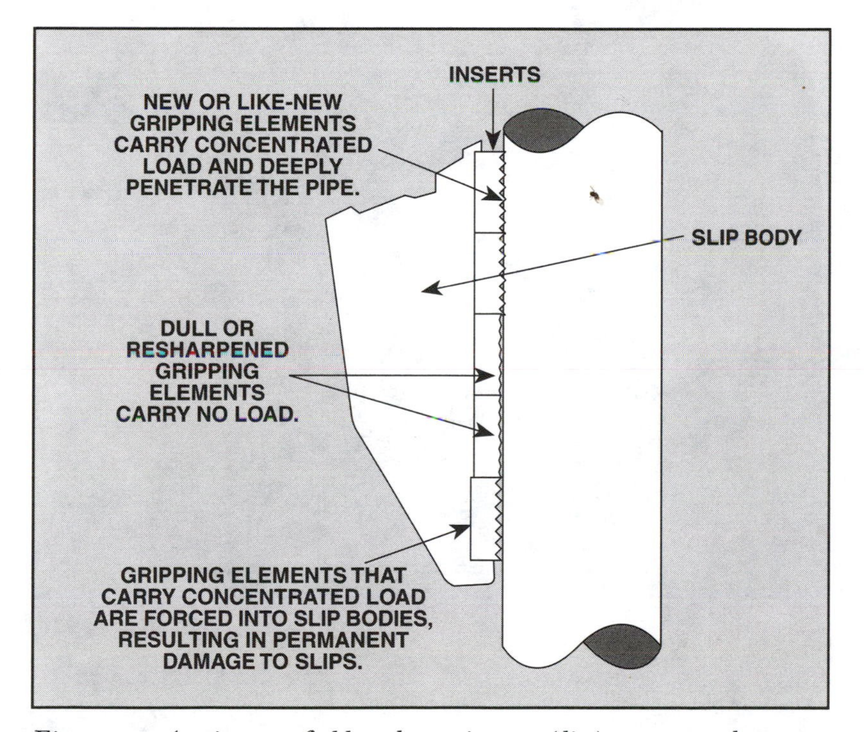

Figure 53. A mixture of old and new inserts (dies) can cause damage.

Maintenance and Inspection

Routine maintenance of slips and master bushings helps ensure a long life. It also protects tubular goods. Proper care consists of—

1. cleaning and lubricating (fig. 54),
2. replacing worn parts (fig. 55),
3. inspecting slips regularly; use a straightedge to check the front and back of the slips and the inserts (dies) for uneven wear or damage (fig. 56), and
4. running slip tests on a regular basis. Slip tests detect wear of the slip inserts and the master bushing. Rotary helpers should do a slip test every three months. They should also run a slip test each time they install a new master bushing or put a new set of slips into service. The test involves imprinting the dies onto a sheet of heavy-weight paper. If the paper shows insert contact on the top section only, this indicates a worn master bushing or worn slips.

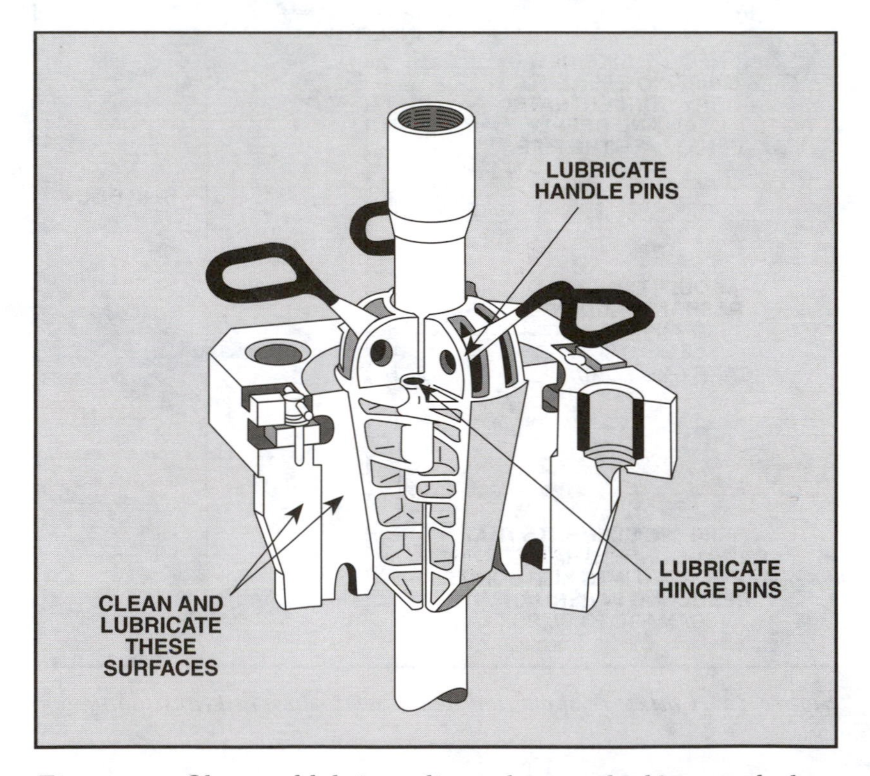

Figure 54. Clean and lubricate slips and master bushing regularly.

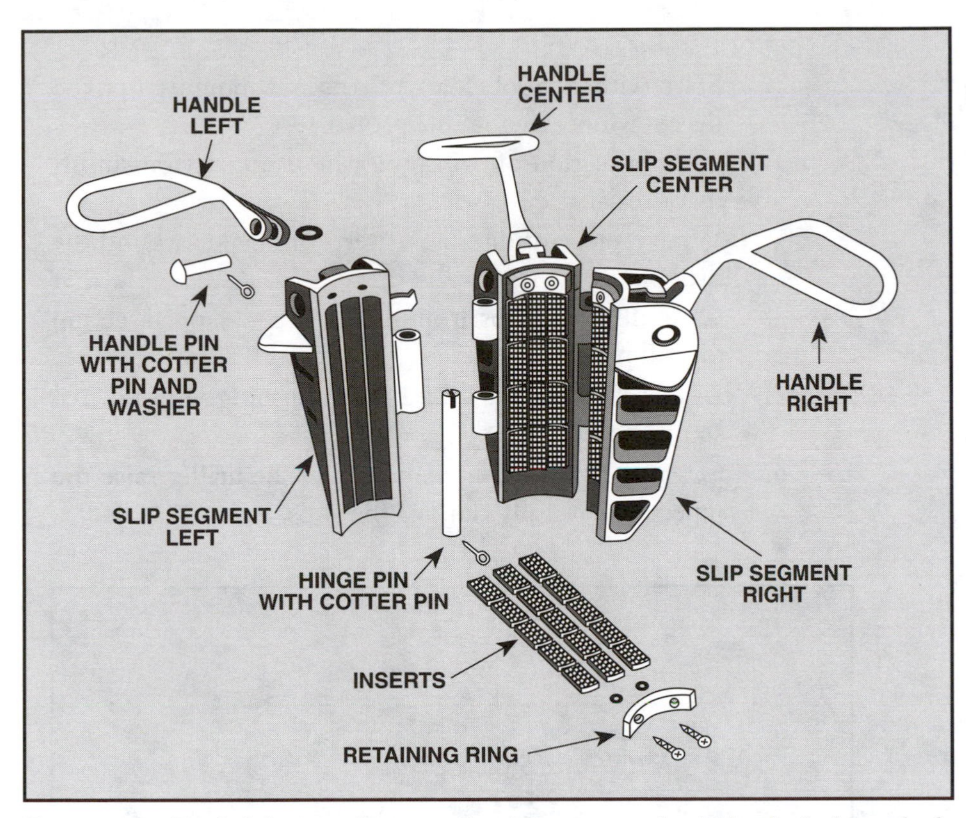

Figure 55. Exploded view of rotary slips. These parts should be checked regularly for wear.

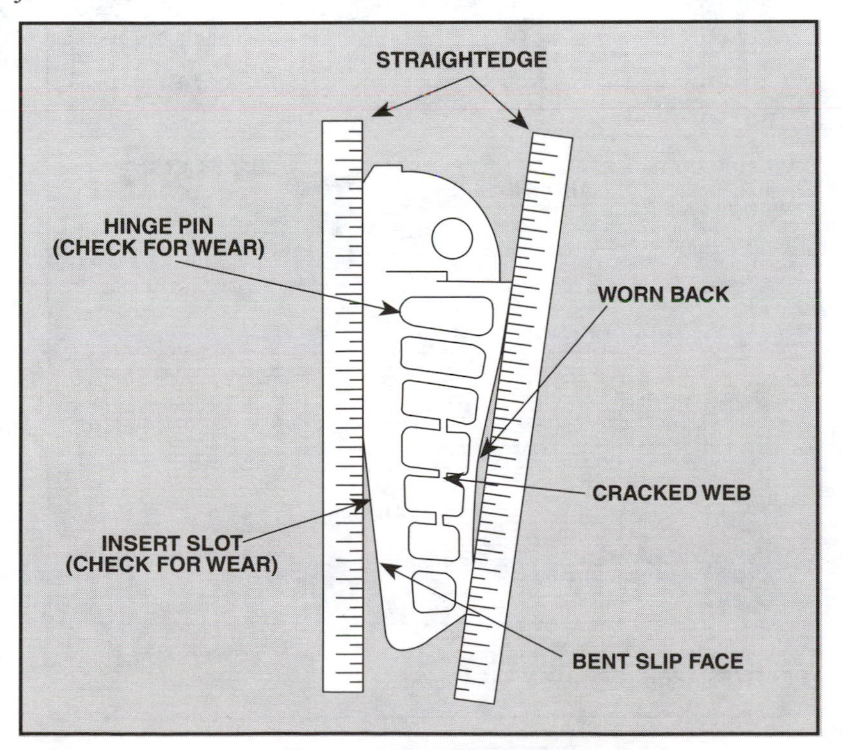

Figure 56. A straightedge is used to check the front and back of the slips.

To perform a slip test (fig. 57)—

1. Start with a hook load of 100,000 pounds (44,500 decanewtons) to obtain a proper test.
2. Clean a section of drill pipe where no previous insert marks exist.
3. Wrap a piece of durable waterproof paper around the drill pipe.
4. Carefully place slips around the paper-wrapped section of drill pipe.
5. Lower the slips into the master bushing with normal setting speed.
6. Grasp the slips by the handles, let the driller raise the pipe, and carefully remove the slips.

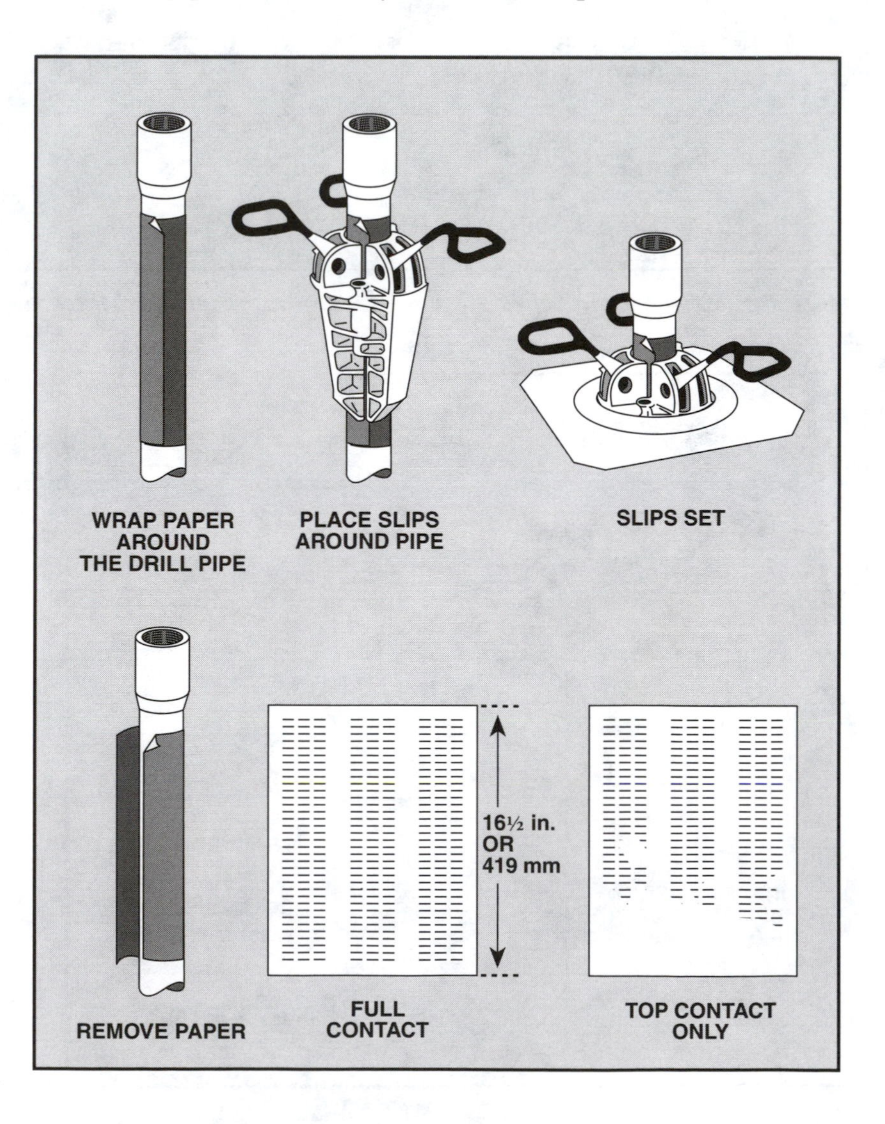

Figure 57. Slip test

7. Remove the paper and evaluate the markings.
   A. If the paper shows full contact, the master bushing and the slips are good, and the test is complete. Full contact, using extra-long slips with a pin-drive master bushing, is 16½ inches (419 millimetres). If the slips do not show full contact, go to the next step.
   B. Rerun the test with new slips. If the result is a full 16½ inches (419 millimetres) of contact, the first slips are worn or damaged and should be discarded. If the paper shows insert contact on the top section only, even with the new slips, replace the insert bowl or the master bushing. Contact only at the top with new slips means one of two things: either the base of the inside taper of the master bushing's bowl is worn or the master bushing itself is worn. In either case, repair or replace the worn part.

Power Slips

Power slips are pneumatic- , hydraulic- , or spring-actuated slips that free rotary helpers from bending to set the slips by hand. The driller controls some models. Others are set by the floor crew's stepping onto an attachment on the outside of the slips. Crew members install the same type of inserts in power slips as they do in manual slips.

To summarize—

Functions of master bushing

- Connects the rotary table to the kelly bushing and transfers rotation from one to the other during drilling
- Holds the slips when drilling stops

Types of master bushing

- Split-construction type
- Solid-construction type
- Hinged-construction type

Types of master bushing drive

- Four-pin
- Square

Functions of the kelly bushing

- Transfers turning motion from the rotary table's master bushing to the kelly
- Allows the kelly to move up and down freely

Kelly bushing design

- Single-plane
- Double-plane

Kelly bushing maintenance

- Lubricate regularly
- Tighten top nuts
- Inspect for wear
- Replace worn parts

Functions of the slips

- Grip the pipe without damaging it
- Suspend the drill stem in the rotary table

Types of slips

- Rotary
- Drill collar
- Casing
- Power

Maintenance of slips

- Clean and lubricate
- Replace worn parts
- Inspect regularly
- Run slip tests on a regular basis

▼
▼
▼

Kelly

Definition

The kelly is a flat-sided, heavy steel pipe that crew members attach to the bottom of the swivel. They attach the other end of the kelly to the drill stem. The kelly bushing and the master bushing transfer the rotary table assembly's rotation to the kelly. The kelly, since crew members make it up on the drill stem, turns the drill stem and bit.

The kelly is usually 40 feet (12 metres) long and has either four or six flattened (not round) sides. Crew members make up several attachments to the kelly. These attachments include the upper kelly cock, the lower kelly cock (drill pipe safety valve), and the kelly saver sub (fig. 58).

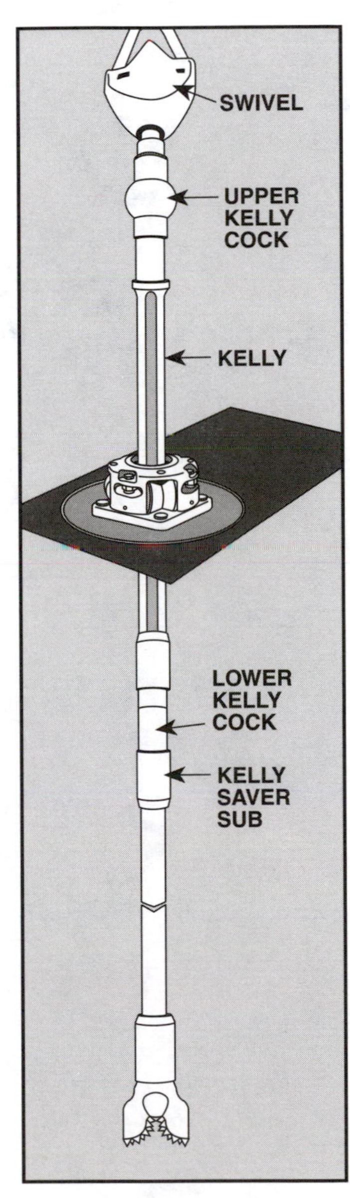

Figure 58. Kelly and attachments

Functions

The kelly—

1. transmits the turning motion of the rotary table to the drill string, and
2. conducts drilling fluid from the swivel to the drill stem.

Design

Kellys are four- or six-sided, instead of round, because the flat sides mate with the kelly bushing to drive (turn) the kelly. A four-sided kelly has a square cross section (fig. 59). A six-sided kelly has a hexagonal cross section (fig. 60). In general, a hexagonal kelly is stronger than a square kelly, but a square kelly is less expensive. As a result, large rigs drilling deep wells often use hexagonal kellys because of their extra strength. On the other hand, small rigs often use square kellys.

The API publishes standards for kellys (API Spec 7K, *Specification for Drilling Equipment*). A standard kelly is 40 feet (12 metres) long, although one 54 feet (16.5 metres) long is also available. A rig drilling

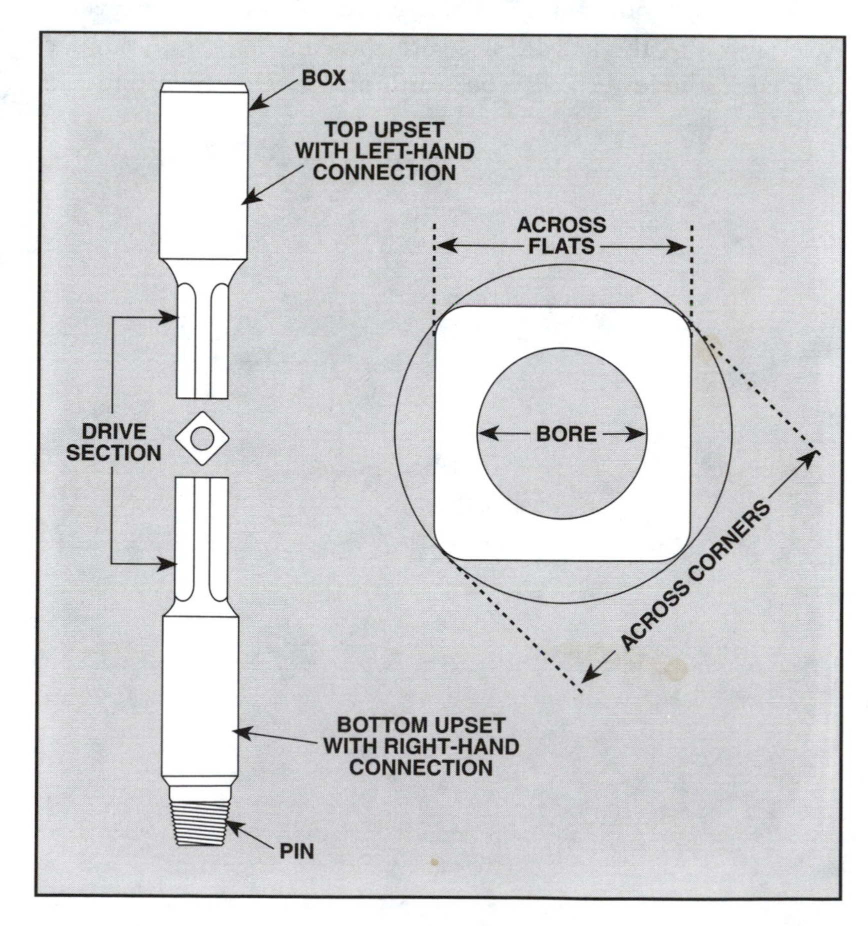

Figure 59. Four-sided (square) kelly

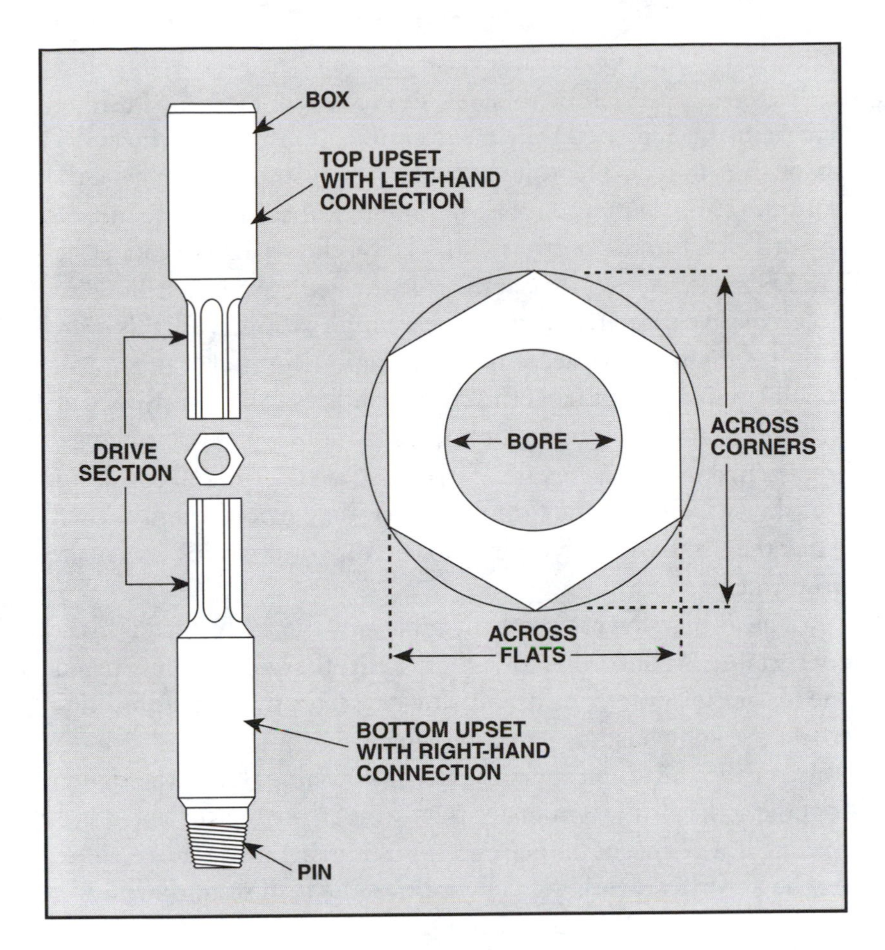

Figure 60. Six-sided (hexagonal) kelly

with range 3 drill pipe (38 to 45 feet, or 11.6 to 13.7 metres, long) could use a 54-foot kelly. Rig owners sometimes use range 3 pipe on rigs with very tall derricks. Most rigs, however, drill with range 2 pipe, which is 27 to 30 feet (8.2 to 9.1 metres) long. Thus, most rigs use 40-foot kellys.

The parts of a kelly include a left-hand box connection, which connects to the swivel, and a right-hand pin connection, which connects to the top joint of drill pipe. Floorhands rotate a left-hand connection to the left (counterclockwise) to tighten it. They rotate a right-hand connection to the right (clockwise) to tighten it. The left-hand connection to the swivel keeps the kelly from unscrewing from the swivel as the drill stem rotates to the right.

Each end of the kelly has an upset. Upsets are thick portions of the kelly into which the manufacturer machines the pin threads and the box threads. The square or hexagonal portion of the kelly is the drive section.

How the Kelly Works

The master bushing and the kelly bushing drive the kelly. Since kellys come with four or six sides, the rollers in the kelly bushing, where the bushing and kelly make contact, must match the kelly's shape (see fig. 33). The rollers inside the kelly bushing engage the flat sides of the kelly and rotate it. Since the drilling crew connects the drill stem to the bottom of the kelly, the stem also rotates.

Let's take a closer look at the role the kelly plays in drilling the hole. A conventional rotary rig (one without a top drive) drills every well, no matter how deep, one drill pipe length at a time. As a starting point, assume that the crew members have just tripped all the drill stem back into the hole after installing a new bit. To get back to making hole, they first retrieve the kelly, the kelly bushing, and the swivel from the rathole, where they placed them earlier. They then stab the kelly's pin into the top joint of pipe and make up the kelly.

The driller lowers the drill pipe and the kelly through the rotary table. As the driller lowers the kelly, the kelly bushing meets the master bushing. The driller slowly rotates the master bushing to get the kelly bushing's pins or square bottom to mate with the holes or the square recess in the master bushing. Once the two fit together, the driller gradually puts weight on the bit and starts making hole. Most of the kelly's length remains above the rig floor. As the bit drills ahead, the kelly follows the drill stem downward, moving through the kelly bushing.

Soon the kelly drills down to a point just above the kelly bushing. At this point, the driller stops the rotary table. The driller then raises the kelly until the top joint of drill pipe appears in the rotary table. The bit is now a kelly's length off bottom. The floorhands set the slips around the top joint of drill pipe and break out the kelly. They then swing the kelly (and attached kelly bushing) over to a joint of pipe in the mousehole. They stab the kelly into the joint and make the two up.

The driller picks up the kelly, the kelly bushing, and the new joint of pipe from the mousehole. Crew members guide the assembly back to the rotary table, where the slips suspend the drill stem. They then stab the new joint into the drill stem hanging in the rotary and make it up tight. The driller picks up the assembly, crew members pull the slips, and the driller lowers the new joint until the kelly bushing engages the master bushing. The driller begins rotating and gradually lowers the kelly until the bit is on bottom again and drilling.

Kelly Accessories

Kelly Saver Sub

The kelly sub, or kelly saver sub, is a short threaded pipe that fits below the kelly (fig. 61). Rotary helpers normally connect the saver sub to the lower kelly cock (see fig. 58). The saver sub minimizes wear on the kelly's threads. Crew members break out the kelly from the drill stem frequently. What is more, they eventually connect the kelly to every joint of drill pipe as the hole gets deeper. If crew members damaged the kelly threads, these threads would damage every thread in the drill stem they came into contact with. To prevent such damage, crew members attach a kelly saver sub to the kelly to protect it and the drill stem. If the kelly saver sub's threads wear or become damaged, crew members simply replace the sub. The sub is a lot less expensive to replace than a damaged kelly. It is also considerably less expensive than the drill pipe.

Frequently, crew members install a rubber protector on the kelly sub. This protector keeps the kelly from rubbing against the inside of the blowout preventers and the top of the casing, which could damage the kelly, the preventers, and the casing.

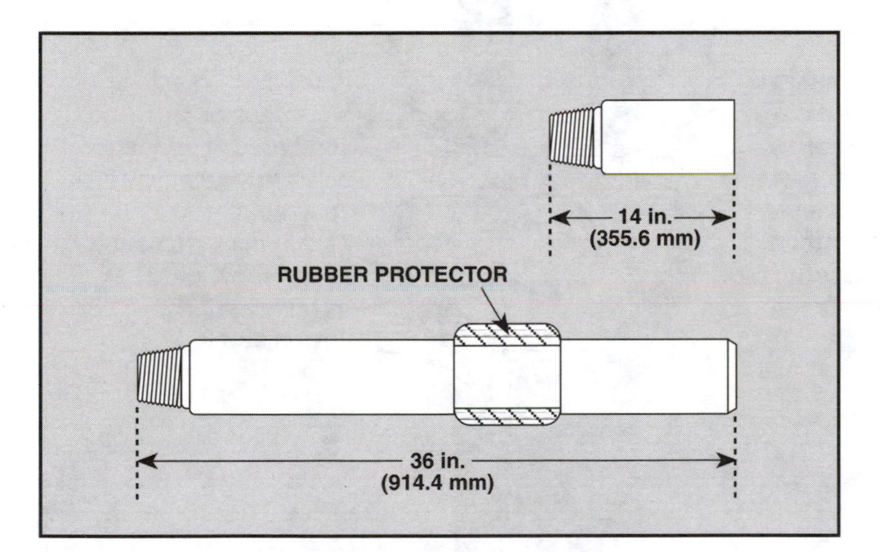

Figure 61. Kelly saver subs

Upper Kelly Cock

Rotary helpers often install a special valve, an upper kelly cock, between the swivel and the kelly (fig. 62). When closed, the upper kelly cock stops the flow of fluids up the drill stem and protects the swivel in emergencies. Sometimes, the hole encounters a high-pressure underground formation. When this happens, formation fluids can enter the well and cause a kick. The kick (the high-pressure fluids) can force drilling mud to flow up the drill stem and escape, or blow out, to the atmosphere. When high-pressure fluids flow up the drill stem, a rotary helper can close the upper kelly cock to keep the pressure off the swivel and rotary hose. Most kelly cocks require a special operating wrench to open or close.

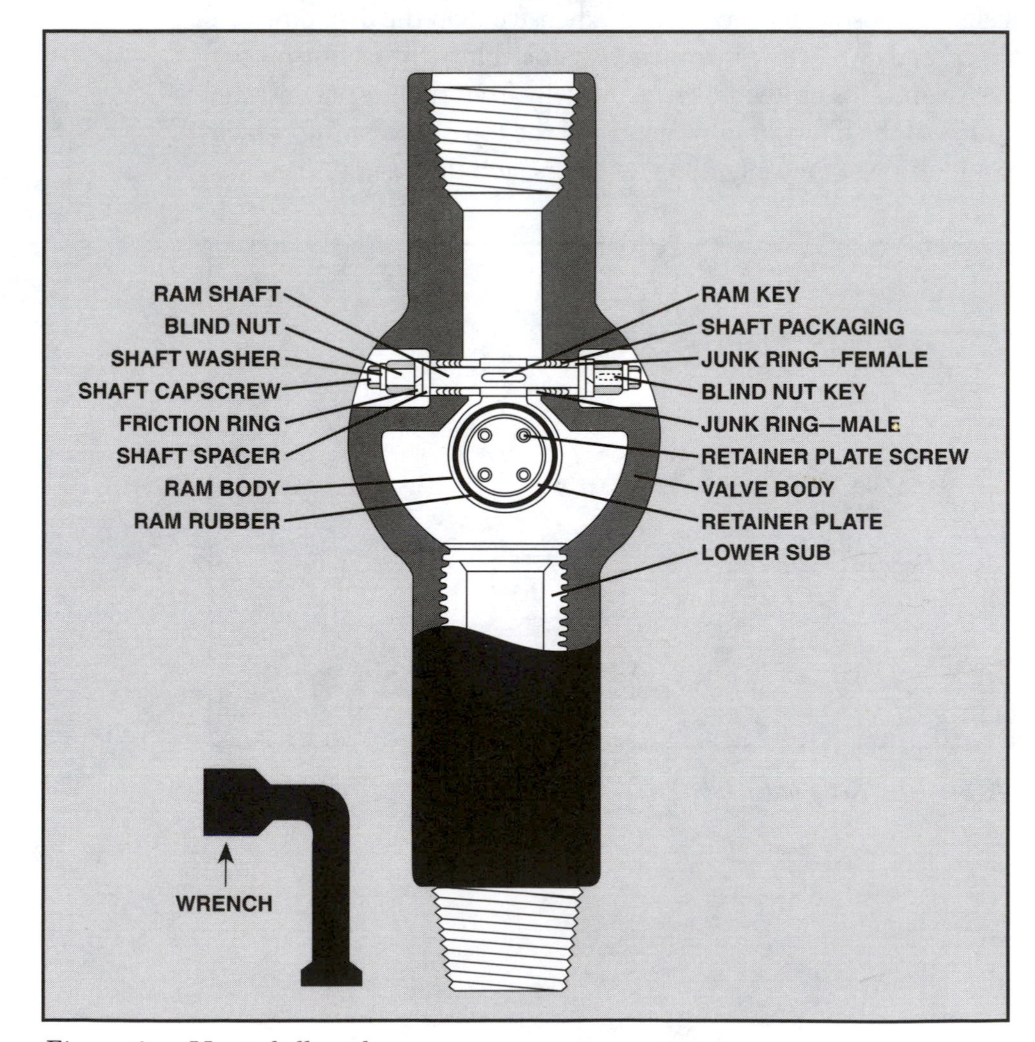

Figure 62. Upper kelly cock

Lower Kelly Cock

A lower kelly cock (a drill pipe safety valve) is, like the upper kelly cock, a standby safety device for blowout prevention (fig. 63). Floorhands routinely make one up below the kelly. Using a special wrench, they can close the lower kelly cock to keep a kick from coming back up through the drill string.

Crew members install a lower kelly cock as well as an upper kelly cock because of accessibility. Suppose, for example, that the kelly is made up on the drill stem and the kelly is drilled down fairly close to the rig floor. A crew member can, in this case, easily close the upper kelly cock if necessary. If, however, the kelly is not drilled down, the upper kelly cock could be 40 feet (12 metres) or so above the rig floor. In such a case, crew members cannot quickly reach it. The lower kelly cock, however, is easy to reach and close.

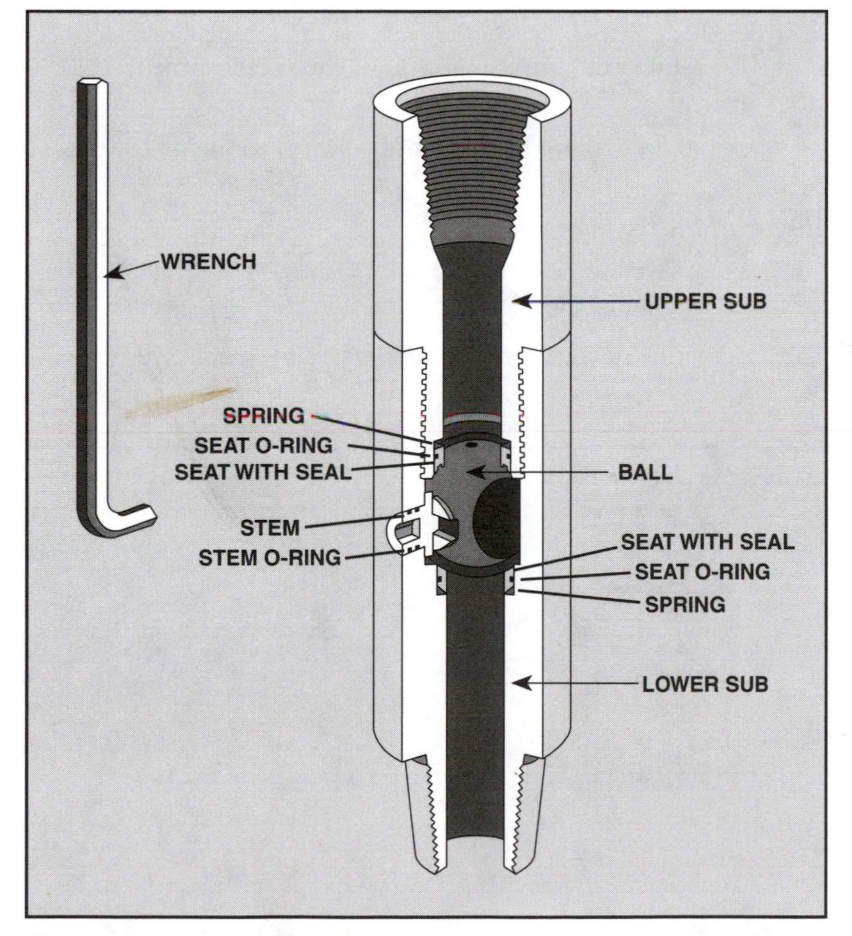

Figure 63. Lower kelly cock

The rotary helpers can also use the lower kelly cock during normal drilling to keep drilling mud from spilling out of the kelly and onto the floor. Crew members therefore sometimes refer to the lower kelly cock as the mud saver valve. When they are ready to make a connection, the floorhands break the kelly out of the drill stem. By closing the lower kelly cock before they break out the kelly, they keep the drilling mud in the kelly. After shutting the valve, crew members should always remember to open it again before circulating through the kelly.

To summarize—

Functions of the kelly—

- Transmits the turning motion of the rotary table to the drill string
- Conducts drilling fluid from the swivel to the drill stem

Attachments to the kelly

- Kelly saver sub
- Upper kelly cock
- Lower kelly cock (drill pipe safety valve, mud saver valve)

Swivel

▼
▼
▼

Definition

Standard dictionaries define a swivel as a device joining two parts so that one or both can pivot freely. The rotary drilling swivel does precisely that. On conventional rotary drilling rigs, the swivel hangs from the traveling block and hook. Crew members attach one end of a strong, steel-reinforced flexible hose—the rotary, or kelly, hose—to the standpipe. (The standpipe is a steel pipe that runs from the mud pump's discharge line and up one leg of the derrick.) They attach the other end of the rotary hose to a short, curved steel pipe, called the gooseneck, which is on the swivel. A passageway inside the swivel conducts drilling mud from the gooseneck to the kelly and drill string. At the same time, the swivel allows the drill stem to turn, or rotate, freely (fig. 64).

Figure 64. Traveling block, hook, and swivel

Function

The swivel—

1. supports the weight of the drill stem during drilling,
2. permits the drill stem to rotate, and
3. provides a passageway for the drilling mud to get into the drill stem.

The swivel can rotate at more than 200 rpm; it can support loads weighing hundreds of tons (tonnes); and it can accommodate drilling fluid pressures greater than 4,000 psi (27,500 kilopascals). The swivel weighs only half as much as the rotary table assembly, but it supports the same load and rotates at the same speed. What is more, a tremendous volume of drilling fluid under very high pressure flows through it.

Design

Manufacturers design swivels for strength and durability and rely on high-test steel. A reinforced cast-steel housing encloses internal machinery and reservoirs for an oil bath. Bail pins on the swivel attach the forged-steel bail to the housing. The bail is similar to a large handle on a bucket. Crew members insert the swivel's bail into the hook on the bottom of the traveling block.

Inside the swivel, basic parts include the gooseneck, an erosion-resistant nozzle that connects to the rotary hose, and the washpipe, which directs mud from the gooseneck to the swivel's stem (fig. 65). The stem rotates freely on bearings inside the swivel and attaches to the kelly. As the stem rotates, mud passes through it and into the kelly and drill stem. The washpipe housing sits above the main body and covers the washpipe packing and an oil seal. The packing and oil seal keep the connection between the washpipe and the stem leakproof. Radial and thrust roller bearings near the top and bottom of the stem center it precisely inside the housing and absorb upward thrust. Main thrust bearings, located near the center of the swivel, carry the full load of the drill stem.

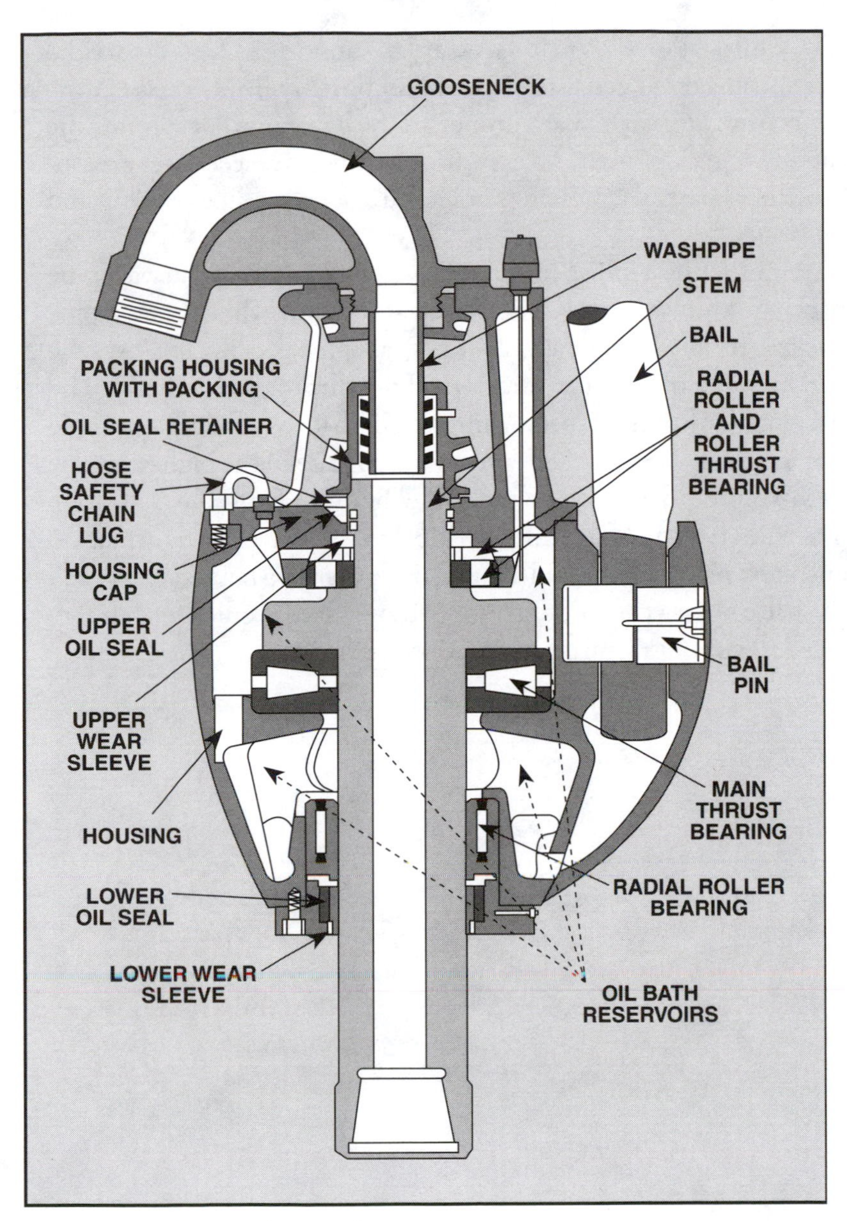

Figure 65. Parts of a swivel

A lubrication system is necessary to support the heavy work the swivel must do. It consists of a combination of oil and grease lubrication (fig. 66). An oil bath immerses the bearings that support the rapidly rotating stem. Grease fittings provide access to grease-lubricated parts and bearings, including the washpipe packing and oil seals.

Another part indispensable to swivel operation is the washpipe packing. The washpipe is locked against rotation, while the washpipe packing turns with the swivel body. Washpipe packing encloses the seal between the static washpipe and the turning swivel body. The packing reinforces this seal against the high-pressure, high-volume drilling mud. The seal thus ensures the passage of the high-velocity, high-volume mud through the swivel body and into the drill stem. The integrity of the packing is critical to the swivel's performance, and crew members should change it regularly. Manufacturers sell the packing in a cartridge assembly that makes it quick and easy to replace without disturbing the gooseneck.

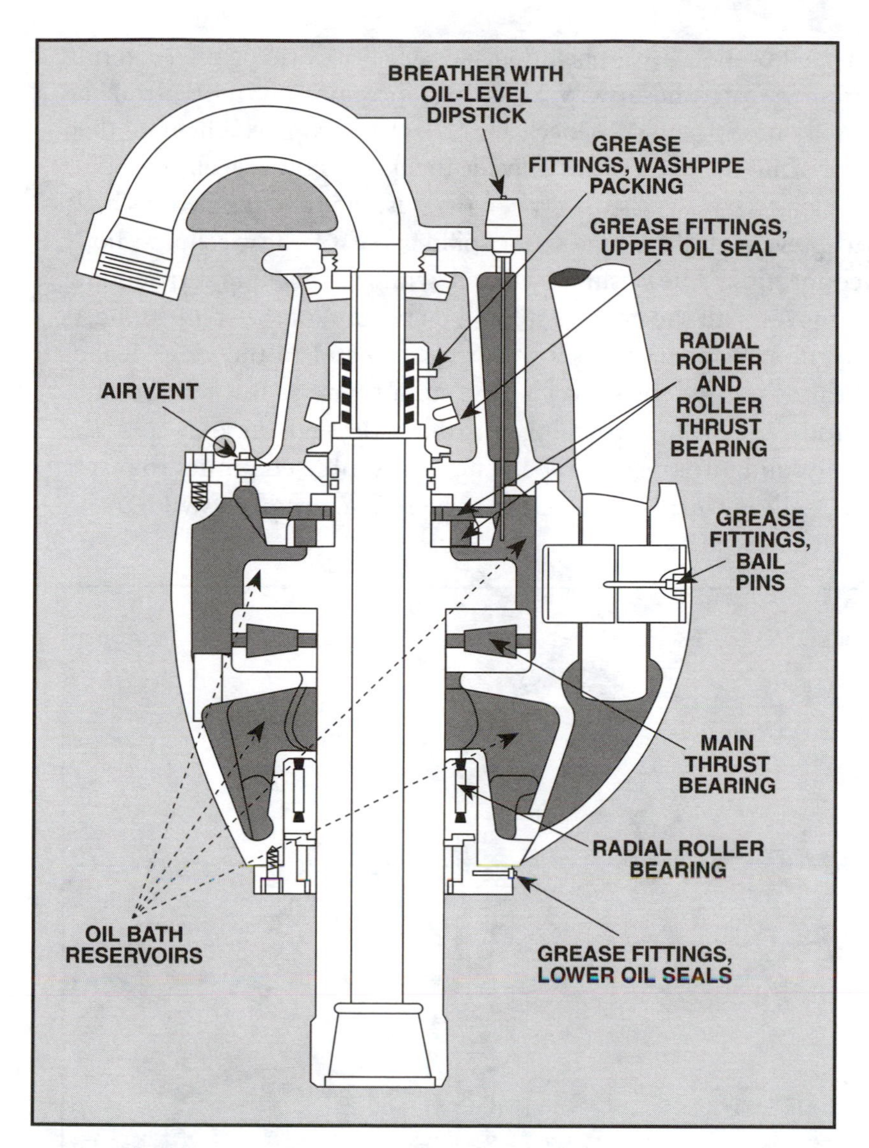

Figure 66. Swivel lubrication system

How the Swivel Works

To follow the swivel's operation, let's start with drilling mud entering from the rotary hose (fig. 67). The high-pressure mud moves from the rotary hose to the gooseneck and into the washpipe. The mud then flows into the stem, which is inside the body of the swivel. Washpipe packing secures the seal between the washpipe and the swivel's body to prevent drilling fluid from spilling out of the washpipe-body connection. The washpipe is fixed, while the packing surrounding it moves with the swivel's stem. The bottom of the stem connects to the kelly cock (not shown in fig. 67) and to the kelly. Roller bearings inside the swivel let the stem rotate as the kelly rotates. Mud passes from the turning stem down through the kelly cock, the kelly, and into the drill pipe. The rotary table continues to rotate the kelly and drill stem, and the swivel stem turns freely with them.

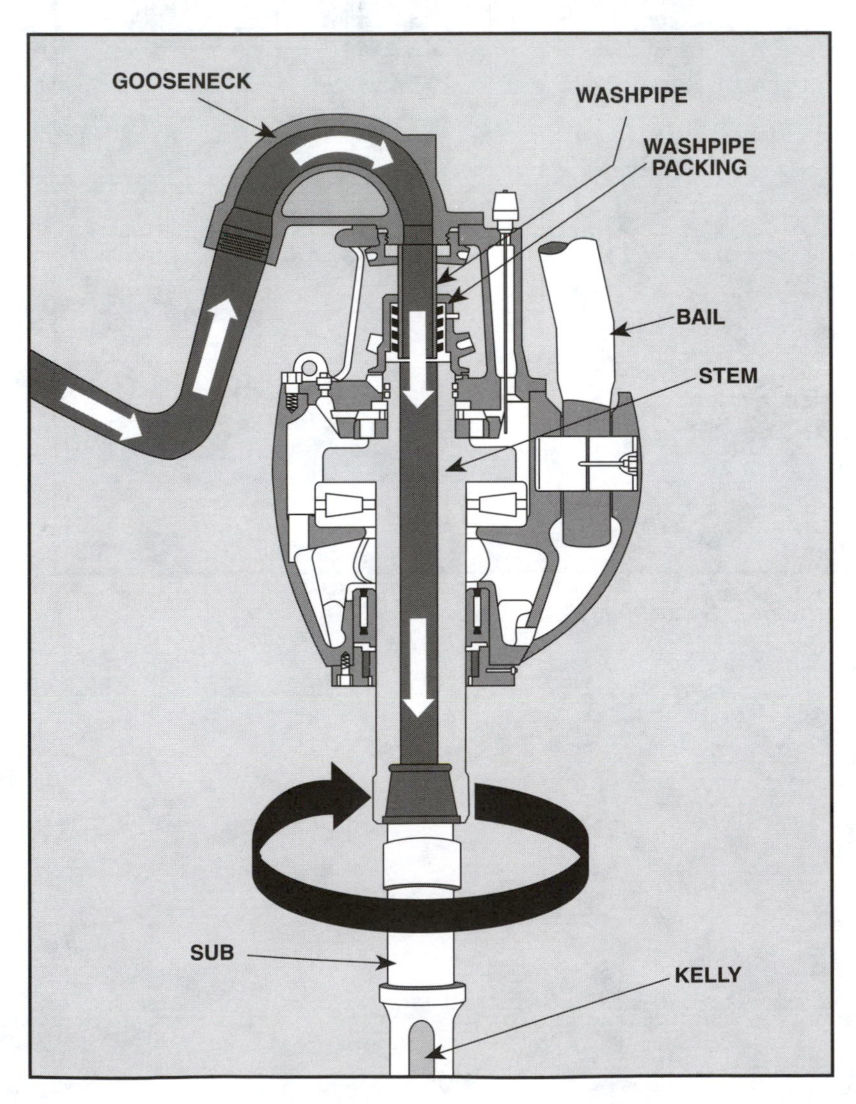

Figure 67. Fluid circulation through the swivel

Maintenance

Recommended swivel maintenance emphasizes (1) regular lubrication; (2) frequent visual inspection for wear, cracks, and leaks; and (3) prompt attention to repairs.

Before changing parts or making repairs, always check the manufacturer's guidelines because the make and model of swivels vary and so does the required maintenance. Manufacturer's manuals contain information on such practices as putting a new swivel into operation, changing the oil, and changing the washpipe packing. Although specific maintenance points may be different, general maintenance for all makes and models is similar.

In general:

1. Keep the swivel clean.
2. Coat the bail throat (the curve in the bail where it hangs from the hook) with grease.
3. Lubricate the bail pins, the oil seals, the upper bearing, and the packing.
4. Check the oil level as recommended by the manufacturer.
5. Change the oil at intervals recommended by the manufacturer.
6. Remove rust and apply weather protection as required.
7. Check and secure all the fasteners.

Replacing Oil Seals and Wear Sleeves

Manufacturers fit swivels with one oil seal at the top and two oil seals at the bottom (see fig. 65). These seals retain lubricant in the housing. The upper oil seal has a replaceable wear sleeve pressed to a shoulder on the body. Check this oil seal and sleeve for wear when you remove the washpipe and packing housing for inspection. The manufacturer seats the bottom two oil seals and their spacers in a retainer bolted to the lower end of the housing. To check these seals, drain the oil from the housing, and remove the retainer. Replace the seals if they are worn or cracked.

Replacing Worn Washpipe Packing

Washpipes and washpipe packing often leak (fig. 68), and replacing them is a frequent maintenance chore. Pressurized, abrasive drilling mud puts them under intense wear. As a result, inspection of these components is critical. If the washpipe or the packing leaks, crew members must replace the damaged part immediately.

Most swivels use washpipe packing that comes as a self-contained cartridge. Washpipe packing kits come with complete cartridges, or with partial replacements. Good practice is to have an entire replacement cartridge on hand as a backup. Normally, floorhands use a partial kit more often than a complete kit because the items that wear the most and therefore need replacing more frequently are the packing and O-rings. When crew members do replace an entire cartridge, they should also repack the old assembly with new packing so it is ready as a replacement. For the procedures on packing removal, inspection, and reinstallation, see Appendix C.

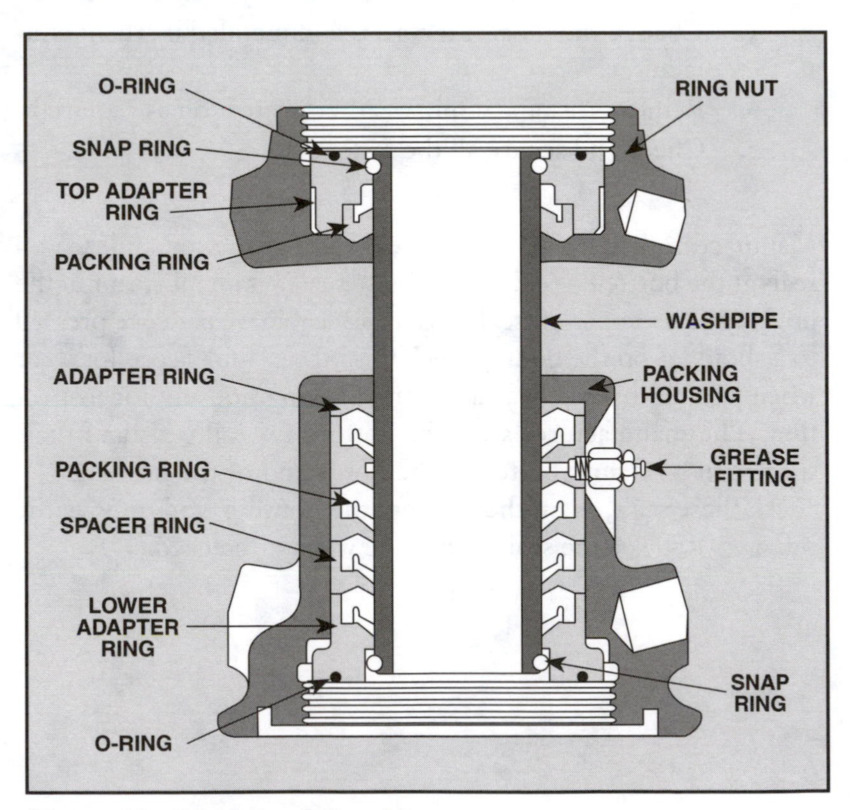

Figure 68. Washpipe and packing

Lubrication

Essential swivel components such as bail pins, bearings, oil seals, and washpipe packing receive heavy wear and depend on lubrication to work (see fig. 66). Always use oil and grease recommended by the manufacturer.

The International Association of Drilling Contractors (IADC) is an organization to which many drilling contractors belong. Among other things, IADC prepares publications to assist rig crew members in doing their jobs better. For example, IADC provides a swivel maintenance checklist (Appendix C is based on this IADC checklist), which drillers and crew members can consult for the proper grades of lubricating oil and grease.

Swivel manufacturers also provide assistance for those who maintain swivels (see Appendix D). For example, most manufacturers agree that crew members should use the following guidelines for swivel lubrication. Every eight hours—

1. Give *all* grease fittings one to three shots of grease.
2. Grease the washpipe packing assembly. To do so, stop the mud pump to relieve pressure on the assembly. With no pressure on the packing assembly, grease can penetrate all its parts.
3. Grease the bail pins while the swivel is set back in the rathole. This at-rest position allows the grease to flow into heavy load–carrying areas.
4. Grease the upper oil seals.

Once a day—

1. Check the oil level while the swivel is set back in the rathole.
2. Screw in the dipstick plug hand-tight.
3. Keep the level at the full mark. Add more oil if necessary.

Every 1,000 hours of operation—

1. Change the oil. Drain the oil from the swivel while it is still warm so that the oil will drain easily.
2. Inspect the magnets in the drain plug. Metal particles clinging to the magnet can indicate excessive bearing wear. If particles are present, schedule the swivel for disassembly and cleaning.

To summarize—

Functions of the swivel

- It supports the weight of the drill stem during drilling
- It permits the drill stem to rotate
- It provides a passageway for the drilling mud to get into the drill stem

Swivel maintenance

- Keep the swivel clean
- Coat the bail throat with grease
- Lubricate the bail pins, the oil seals, the upper bearing, and the packing
- Check the oil level as recommended by the manufacturer
- Change the oil at intervals recommended by the manufacturer
- Remove rust and apply weather protection as required
- Check and secure all fasteners

▼
▼
▼

Spinning and Torquing Devices

Spinning and torquing devices on a conventional rotary rig include large manual wrenches (fig. 69), power wrenches, chains, and other equipment that turn drill pipe. Crew members use this equipment to connect or disconnect the pipe. They frequently remove the drill stem from the hole and disassemble it. Crew members also often join lengths of drill pipe together. For example, they may need to add a length of drill pipe to drill ahead. Or, they may need to trip out the entire drill string to change bits.

When connecting pipe, floorhands use spinning tools first, and then powerful torque wrenches to finish the job. The spinning equipment rapidly rotates, or spins, the joint together. Crew members then use torquing tools to make the joint up to final tightness. Spinning equipment also spins out the joint after the floorhands break the two members of the joint apart with torquing equipment. Examples of spinning tools are the spinning chain, the kelly spinner, and the spinning wrench.

Figure 69. Using manual tongs to tighten a joint

Torquing devices apply the turning force, or torque, needed to make a connection up to final tightness. Floorhands also use torquing devices to break apart the tightly sealed connection between the joints of pipe. Torquing devices include manual tongs and power tongs.

Spinning Devices

Spinning Chain

Definition

The spinning chain is a Y-shaped chain crew members use to spin one joint of drill pipe into another (fig. 70). Crew members attach one end of the chain to the tongs and leave it there until they finish the well and move on to the next hole. Similarly, they attach another end to the makeup cathead on the driller's side of the drawworks. The floorhands leave the third end free. One crew member wraps this free end around a joint of drill pipe that the crew has stabbed into a joint of pipe hanging in the rotary table.

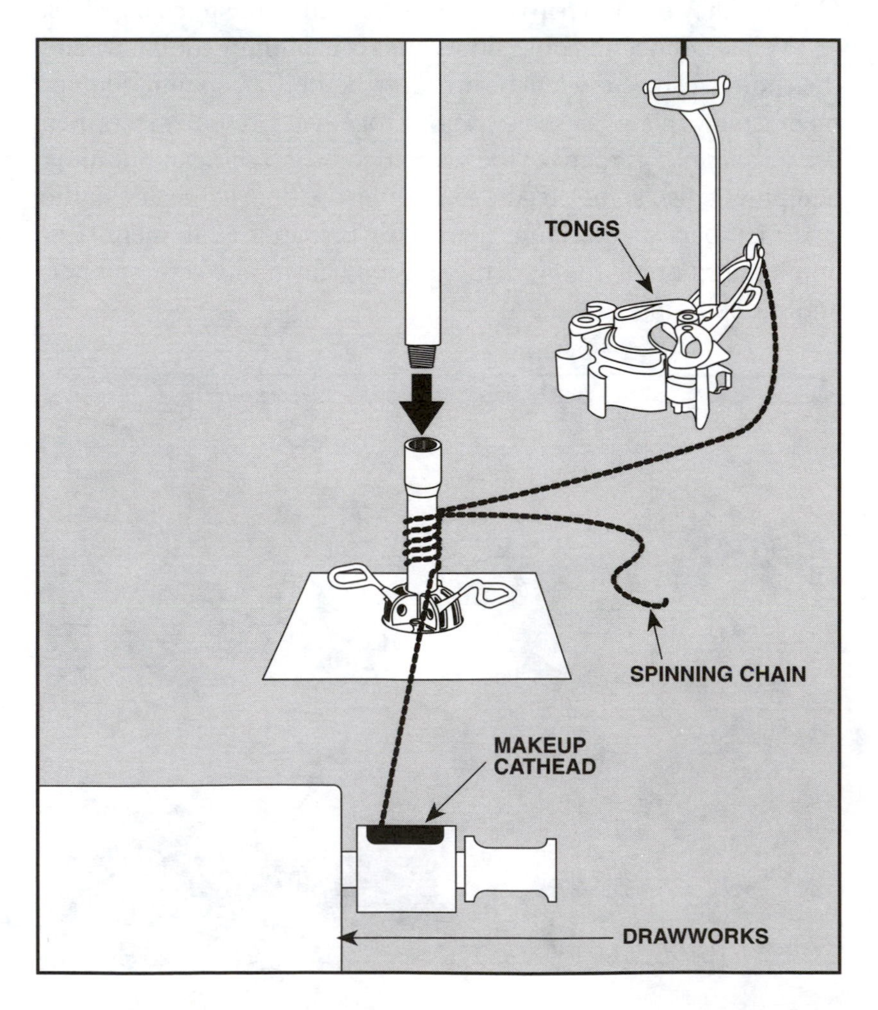

Figure 70. Spinning chain

Function

The floor crew uses the spinning chain in conjunction with cathead-operated tongs. The spinning chain rapidly spins, or rotates, a joint of pipe that crew members stab into a joint of pipe hanging in the rotary table. When the driller engages the makeup cathead, the cathead pulls the chain from around the joint of pipe and causes the pipe's threads to screw up, hand-tight, in the suspended joint. Crew members then use the tongs to make up the pipe to final torque.

How the Spinning Chain Works

Crew members use the spinning chain when tripping pipe back into the hole. To begin, the driller lowers a stand of drill pipe into the hole. When the top stand's tool joint is at the right height in the rotary, crew members set the slips. The female, or box, end of the tool joint is showing. The floorhands then wrap the free end of the spinning chain neatly around the tool joint suspended in the hole. Keep in mind that the second end of the spinning chain is attached to the end of the cathead-operated tongs. The third end runs to the makeup cathead.

The driller, the derrickhand, and the rotary helpers pick up the next stand of pipe out of the derrick and maneuver it over the pipe hanging in the rotary. The floorhands stab the pin of the new stand into the box suspended in the hole. Now one of the rotary helpers "throws the chain." The helper combines a skillful twist of the wrist with an upward flipping motion. The chain uncoils from the pipe in the rotary and coils around the tool joint of the just-stabbed stand. The driller then engages the cathead to pull on the spinning chain. A rotary helper controls the loose end. As the chain comes off the pipe, it causes the stand to rotate clockwise. This motion screws the spinning stand into the suspended stand. Then crew members use the conventional tongs to buck up the stand to final tightness.

Kelly Spinner

A kelly spinner is a hydraulically or pneumatically powered device that spins the kelly and the drill pipe attached to it to thread or unthread connections (fig. 71). Crew members mount it below the swivel and on top of the kelly. It is safer and faster than a spinning chain. It turns the kelly without requiring the floorhands to work with a spinning chain around the rotary table or the mousehole. Speed is a big advantage during a kick when well control requires making up the kelly onto drill pipe.

Figure 71. Kelly spinner

Spinning Wrenches

Spinning wrenches are hydraulic- or air-powered wrenches used to spin drill pipe when making up or breaking out connections (fig. 72). To use spinning wrenches, rotary helpers swing and latch them onto the pipe body above the tool joint. They then actuate the wrench to make it spin up or spin out the pipe.

Spinning wrenches come in several models. Air-driven spinners use the same rig air that powers the air hoist and other rig equipment. Hydraulic spinners require their own power supply. Both fit pipe sizes up to 9 inches (228.6 millimetres) and feature variable speed control. A snubbing or safety line attached to the rig structure keeps the tongs from spinning around the pipe. A wireline anchored in the derrick suspends the wrenches. (See Appendix E for maintenance and lubrication guidelines for one brand of spinning wrench.)

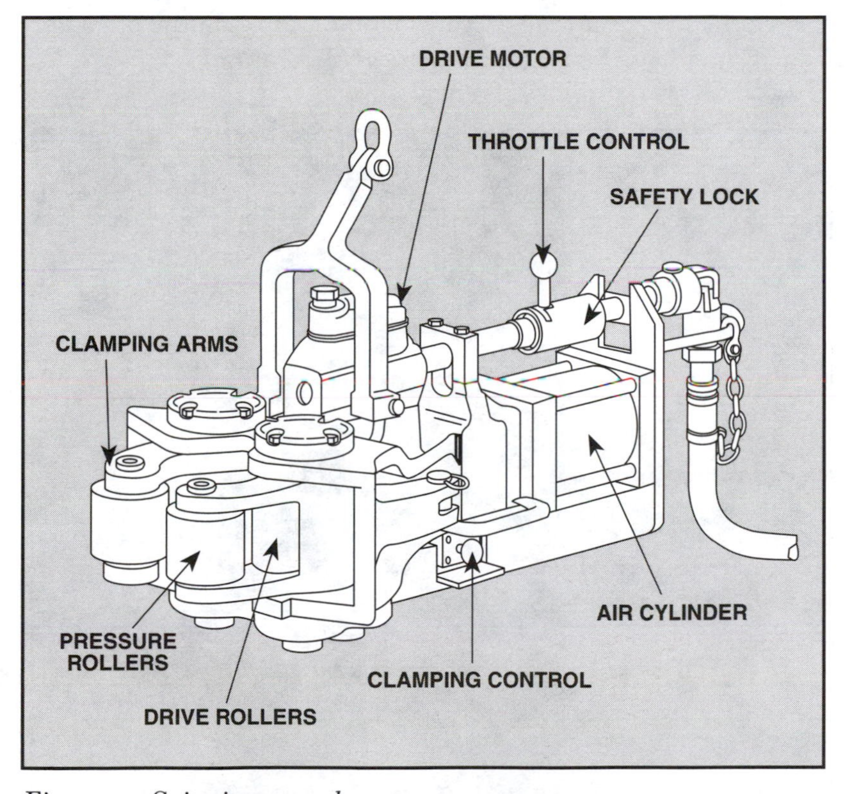

Figure 72. Spinning wrench

Torquing Devices

Conventional Tongs

Definition

Conventional tongs are large, hand-latched wrenches that hang from a wire rope. The wire rope (the tong hanger line) runs from the tong's hanger, up the derrick, over sheaves attached to the derrick, and down to a counterbalance (fig. 73). The counterbalance may be in the derrick or beneath the rig floor. The counterbalance offsets the weight of the tongs to keep them suspended at a convenient working height above the rig floor.

Rig crews use two sets of tongs to make up and break out drill pipe. When making up or breaking out drill pipe, they latch one set of tongs on the tool joint of the pipe suspended in the rotary table. This set is the backup tongs, and they keep the pipe from turning in the slips. Crew members latch the other set of tongs on the tool joint to be made up or broken out. If they are making up pipe, they call this set the makeup tongs; if they are breaking out pipe, they call this set the breakout tongs. Crew members also call the breakout or makeup tongs the lead (pronounced "leed") tongs.

Figure 73. Tong counterbalance in derrick

Function

The rotary crew uses tongs to make up pipe connections to final tightness and to break out the tightly sealed connections. They also use a set to back up the pipe—that is, to keep the pipe from turning in the slips as the makeup or breakout tongs apply a large amount of torque.

Crew members also use tongs to make up and break out casing, tubing, or other pipe. In such cases, they call them casing tongs, tubing tongs, and so forth, according to the specific use.

Design

Tongs feature a hanger, jaws, inserts (or dies), and a handle or arm (fig. 74). The crew matches the tong jaws to the size of the pipe or the drill collar by installing the correct size jaws for each size of pipe. This provides a snug grip and also prevents damage to the drill pipe. Tong inserts, or dies, are hard, brittle pieces of serrated steel that grip (bite into) the tool joint when crew members latch the tongs onto the pipe (fig. 75). They eventually wear out, but floorhands can easily replace them by sliding them out of slots in the tong jaws and then sliding a replacement into the same slot. Even if only one set of dies is worn, most manufacturers recommend replacing all of them.

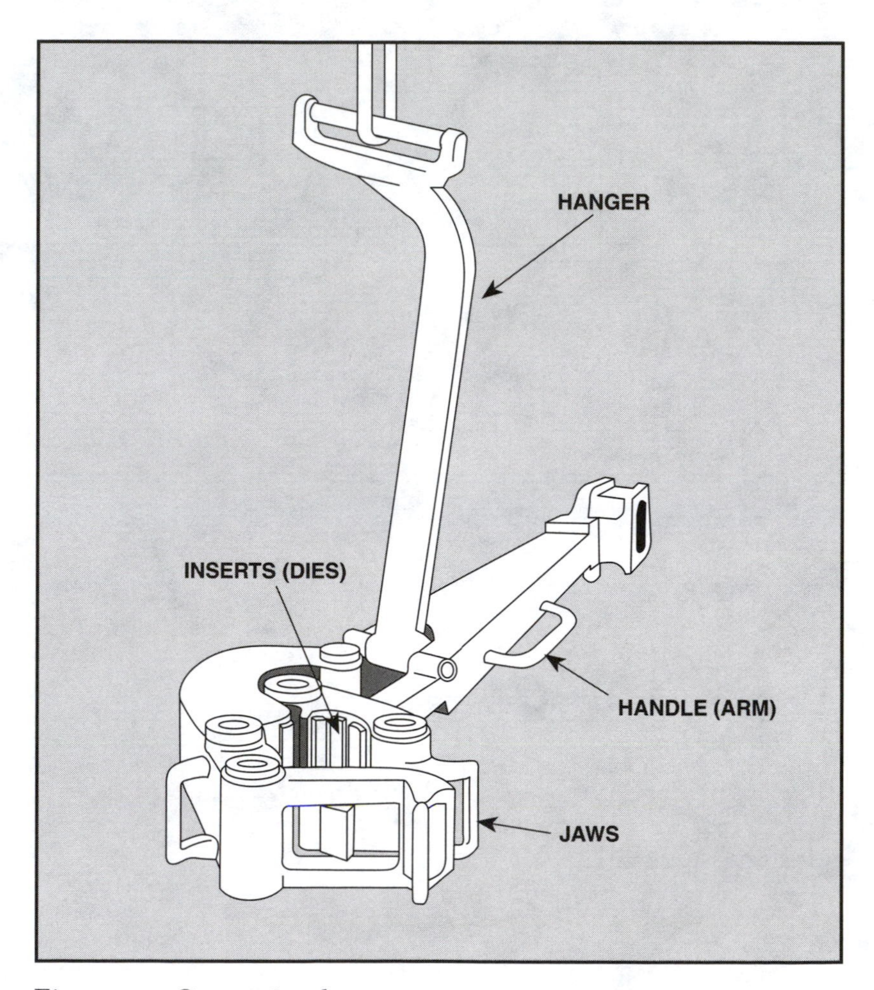

Figure 74. Conventional tongs

Figure 75. Tong inserts (dies)

Crew members attach one end of a wire rope to the hanger and the other end to the tong counterbalance. This wire rope suspends the tongs above the rig floor.

Another wire rope—the snub line—runs from a derrick leg to the end of the tong handle. This line keeps the tongs from turning too far when the tongs are making up or breaking out pipe. The spinning chain on the end of the makeup tongs runs to the makeup cathead on the driller's side of the drawworks. A wire rope, called the jerk line, on the end of the breakout tongs runs to the breakout cathead on the other side of the drawworks.

How Conventional Tongs Work

Floorhands normally use conventional tongs with the spinning chain or the jerk line. Both conventional tongs and the spinning chain are hand-operated by the rotary helpers with a mechanical pull supplied by the automatic cathead on the drawworks.

To make up a joint, crew members first spin it up with the spinning chain. Then, they latch two sets of tongs onto the tool joints of the pipe. Makeup tongs turn the top joint clockwise. Backup tongs hold the bottom joint still. The driller engages the makeup cathead, which pulls on the chain running to the makeup tongs and turns the connection to final torque.

To break out a connection, the crew switches the position of the tongs. Now the breakout tongs turn the top pipe counterclockwise. The backup tong holds the bottom joint still.

Power Tongs

Air or hydraulic fluid drives power tongs. Crew members use power tongs to torque pipe up to final tightness and break out sealed pipe connections (fig. 76). Some models of power tongs require backup tongs. Others require no backup tongs because their design includes a self-contained backup device.

Power tongs offer several advantages. For one thing, crew members can preset and adjust the amount of torque needed to make up a particular joint properly. This adjustment allows the tongs to repeat a required torque automatically on each tool joint. As a result, less reliance is placed on the driller's sense of timing. In contrast, conventional manual tongs rely on the driller to apply the proper amount of power to catheads at precisely the right moment.

Crews often use power tongs in conjunction with spinning wrenches. To break out pipe, power tongs separate the pin of the top joint from the box of the bottom joint. Then rotary helpers latch the spinning wrenches onto the pipe and rotate the pin out of the box. They reverse the procedure when going into the hole. Crew members stab the pin into the box. They latch the spinning wrench on the top joint and operate its spinners to screw the joint to hand-tightness. They then torque the joint up to final tightness with the power tongs, usually to an automatically preset value. In some cases, crew members apply the correct torque by reading the torque indicator and controlling the power manually.

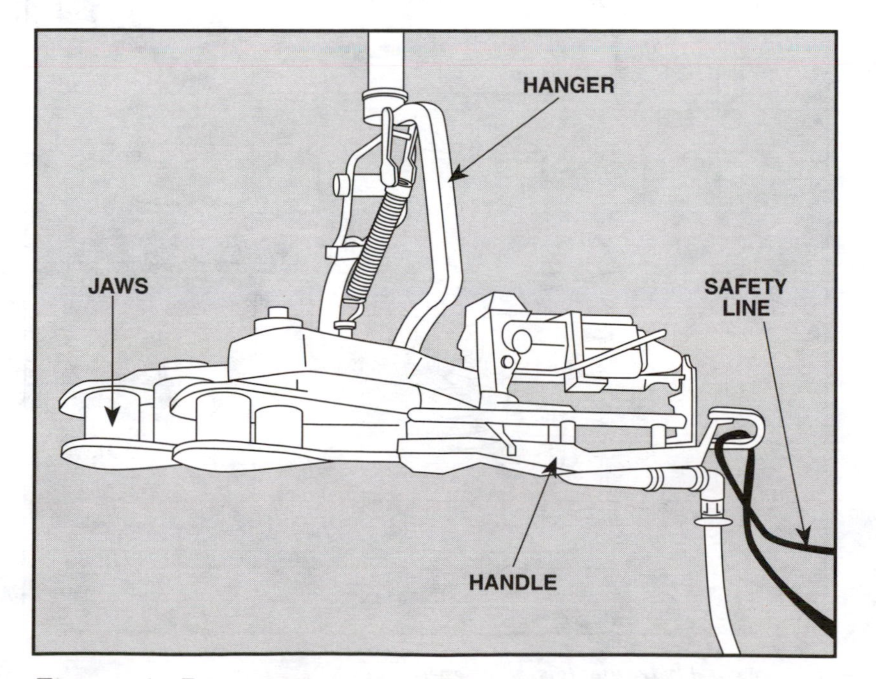

Figure 76. Power tongs

Hydraulic Torque Wrench

The hydraulic torque wrench combines lead and backup tongs in one tool (fig. 77). It also ensures accurate torque. The wrench is fitted with a gauge to measure torque. A repeater gauge duplicates the reading at the driller's console. The driller refers to the repeater gauge to monitor the tool joints as they go in the hole. This procedure double checks all the connections to make sure they have the correct torque. Crew members suspend the torque wrench from the derrick, just as they do conventional tongs.

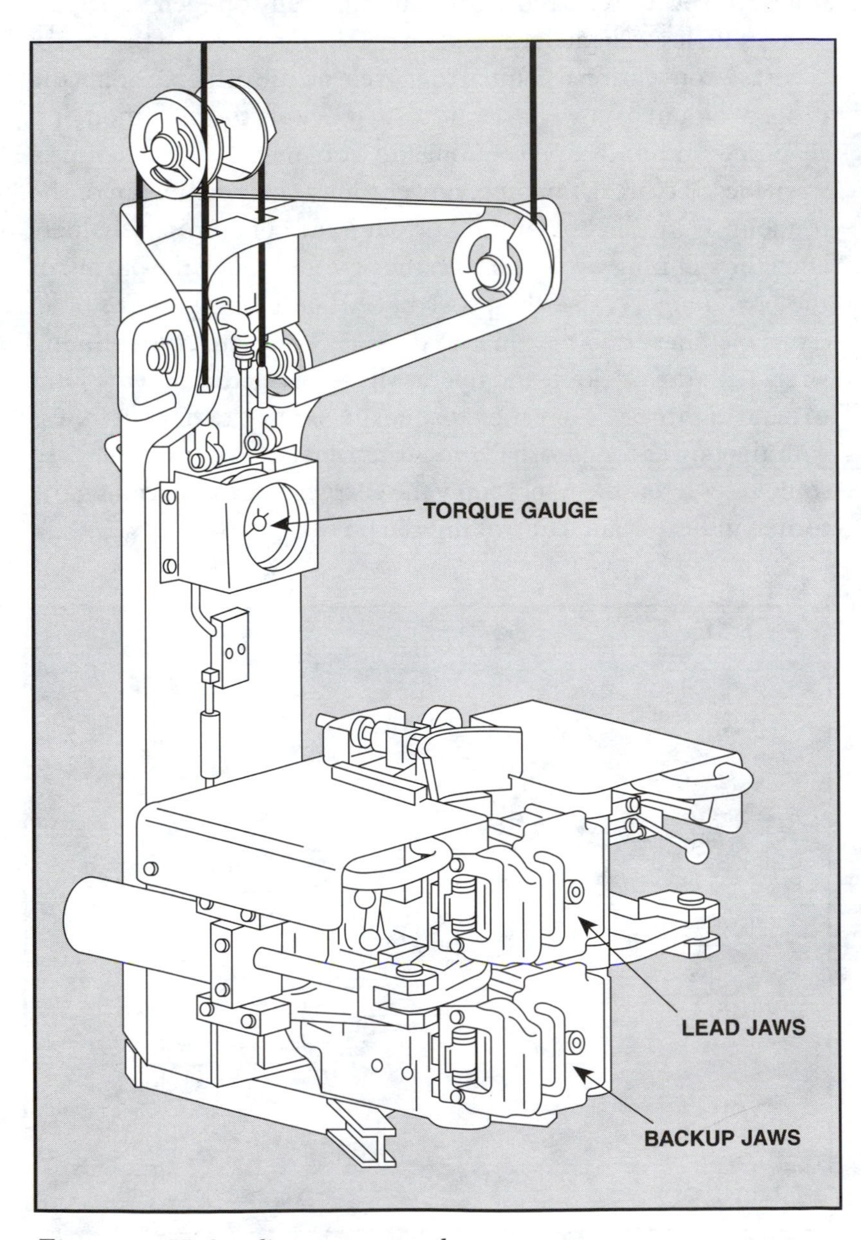

Figure 77. Hydraulic torque wrench

Crew members swing the wrench onto the tool joint. They clamp the tool's backup jaws onto the box, and place the lead jaws onto the pin. Then they activate the tool to rotate the pin to the left to break it out or to the right to make it up. The crew then releases the jaws and swings the torque wrench out of the way. Finally, they use spinning wrenches to spin out or spin up the stand. (See Appendixes F and G for lubrication and maintenance pointers for one brand of torque wrench.)

Combination Power-and-Spinning Wrenches

One power device that saves a lot of time combines spinning tongs and a torque wrench into a single unit. These combination power-and-spinning wrenches are also known by one manufacturer's name, Iron Roughneck™. They mount directly on the rig floor near the rotary table (fig. 78).

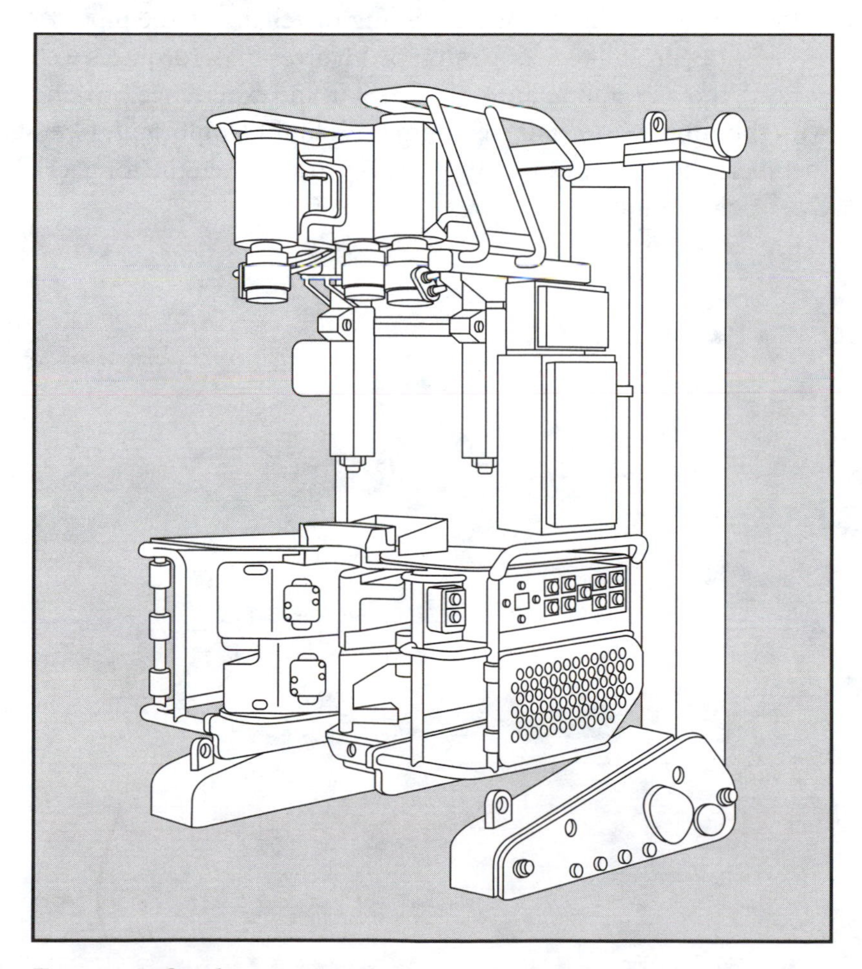

Figure 78. Combination power-and-spinning wrenches (Iron Roughneck™)

Tong Safety

Wire rope suspends tongs from the derrick. It also secures them to the derrick's legs. To ensure that the rope does not become unfastened, crew members always use safety clamps to attach the wire rope. Both the suspension (hanger) line and the safety line (snub line) should be inspected regularly to prevent any failure. The counterbalance for each set of tongs should be placed either under the rig floor or in the derrick using special tracks or a runner. If the counterbalance is suspended in the derrick, it must be secured to a safety line so that it cannot fall to the floor. Tongs should be tied or set back out of the work area when not in use.

Crew members should regularly inspect the spinning chain for large dents or breaks and should replace it immediately if it appears damaged. They should also ensure that the guide line (usually a short piece of soft line) attached to the spinning chain is in good shape. If it is too frayed or ragged, it may be difficult to grip, and a floorhand could have trouble guiding the chain off the pipe.

Making up and breaking out pipe involves high torque as well as heavy moving equipment and pipe. Skill and experience working with spinning and torquing equipment is highly important. Crew members should pay especially close attention when the tong jerk line or chain pulls on the tongs.

Maintenance

Crew members should check the tongs' counterbalancing mechanism regularly and adjust it for proper balance. If crew members have trouble getting the tongs to hang at the correct level, they may add or take away weight from the counterbalance. They should also inspect tong dies regularly and replace them when worn. Floorhands should also grease the tongs before and after each trip and lubricate the tong pins and latch springs frequently. Clean tongs are easier to operate, and they last longer. They should always be cleaned before lubrication so that the grease and oil do not wash off.

To summarize—

Spinning devices

- Spinning chain
- Kelly spinner
- Spinning wrenches

Function of the spinning chain

- Rotates a joint of pipe stabbed into another joint of pipe hanging in the rotary table; used in conjunction with cathead-operated tongs

Function of the kelly spinner

- Spins the kelly attached to the drill pipe to thread or unthread connections

Function of the spinning wrench

- Spins pipe when making up or breaking out connections

Torquing devices

- Conventional tongs
- Power tongs
- Hydraulic torque wrench
- Combination power-and-spinning wrench

Functions of power tongs

- Make up pipe connections to final tightness
- Break out tightly sealed pipe connections
- Back up pipe as makeup or breakout tongs apply torque

Functions of hydraulic torque wrench

- Combine lead and backup tongs
- Ensure accurate torque

Function of combination power-and-spinning wrench

- Combines spinning tongs and torque wrench

Tong maintenance

- Check counterbalance regularly for proper balance
- Inspect tong dies regularly; replace when necessary
- Grease before and after each trip
- Lubricate tong pins and latch springs frequently
- Clean tongs before lubrication

▼
▼
▼

Top Drives

▼
▼
▼

Definition

A top drive is a system suspended in the derrick that works as a kind of power swivel (fig. 79). Modern units combine the elevators, the tongs, the swivel, and the hook.

A top drive performs several rotary drilling jobs at one time:

- it rotates the drill stem;
- it serves as a passageway for drilling mud; and
- it supports the drill stem in the hole.

The rig uses a top drive in place of the regular swivel, the kelly, the kelly bushing, and the rotating function of the rotary table. Even on rigs with a top drive, however, the rig owner retains the rotary table and master bushing as a place for the floorhands to set the slips to suspend the drill stem in the hole.

Figure 79. Top drive in derrick

Functions

The top drive replaces the rotating function of the rotary table assembly, the kelly, the kelly bushing, and the regular swivel. It performs the same jobs that they perform. In addition, a top drive can reduce the number of connections crew members have to make when drilling. By taking the time to make up three-joint stands before drilling commences, rig crews can save time later because they have to make fewer connections when drilling.

Because the top-drive unit has built-in makeup, breakout, and spinning tongs, it saves trip time and is safer. The unit's built-in tongs allow crew members to make up and break out pipe directly from the string. No chain or line runs from the drawworks catheads, which reduces safety hazards. The power unit (the motor) that turns the top drive's drive shaft also powers the tongs.

A top drive allows the driller to ream long sections of hole, which can sometimes help prevent the drill stem from getting stuck. Reaming the hole means rotating the pipe and bit, circulating drilling mud, and moving the pipe up and down. Being able to ream can sometimes help if the drill stem tends to become stuck in the hole. With conventional drilling, the driller can ream a section of hole only equal to the length of the kelly. With a top drive, the driller can ream the hole a distance that is limited only by the height of the derrick, which, on rigs using top drives, is usually over 100 feet (30 metres).

As in a conventional swivel, kelly, and kelly bushing system, crew members hang a top drive from the rig's traveling block and hook. A top drive does not, however, require a kelly and a kelly bushing. Instead, crew members attach the drill string directly to a drive shaft in the unit. A powerful motor (or motors) rotates the drive shaft and attached drill string.

The top drive has a built-in inside blowout preventer (IBOP), which is similar to conventional upper and lower kelly cocks (fig. 80). If a kick occurs during drilling, the driller can close the IBOP by remote control. This prevents the kick from exiting to the atmosphere through the top drive.

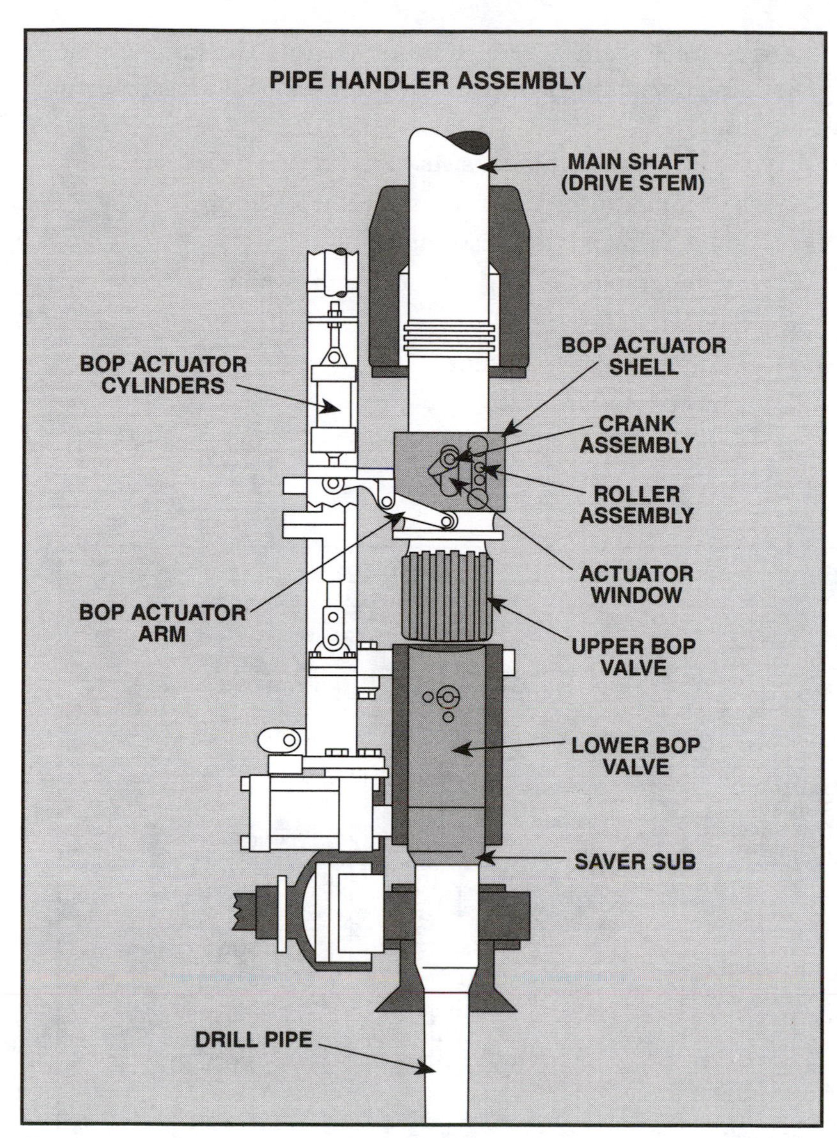

Figure 80. Built-in inside blowout preventer (IBOP) on top drive

Design

Top drives are actually several systems assembled into one unit (fig. 81). Each system performs its own function and has its own subsystems.

A typical top-drive unit consists of—

1. a drilling motor and a transmission assembly
2. guide rails and dolly assembly
3. an integrated swivel gooseneck and S-pipe
4. a pipe handler
5. a counterbalance system
6. a motor cooling system
7. a control system (electrical service loop and control panel).

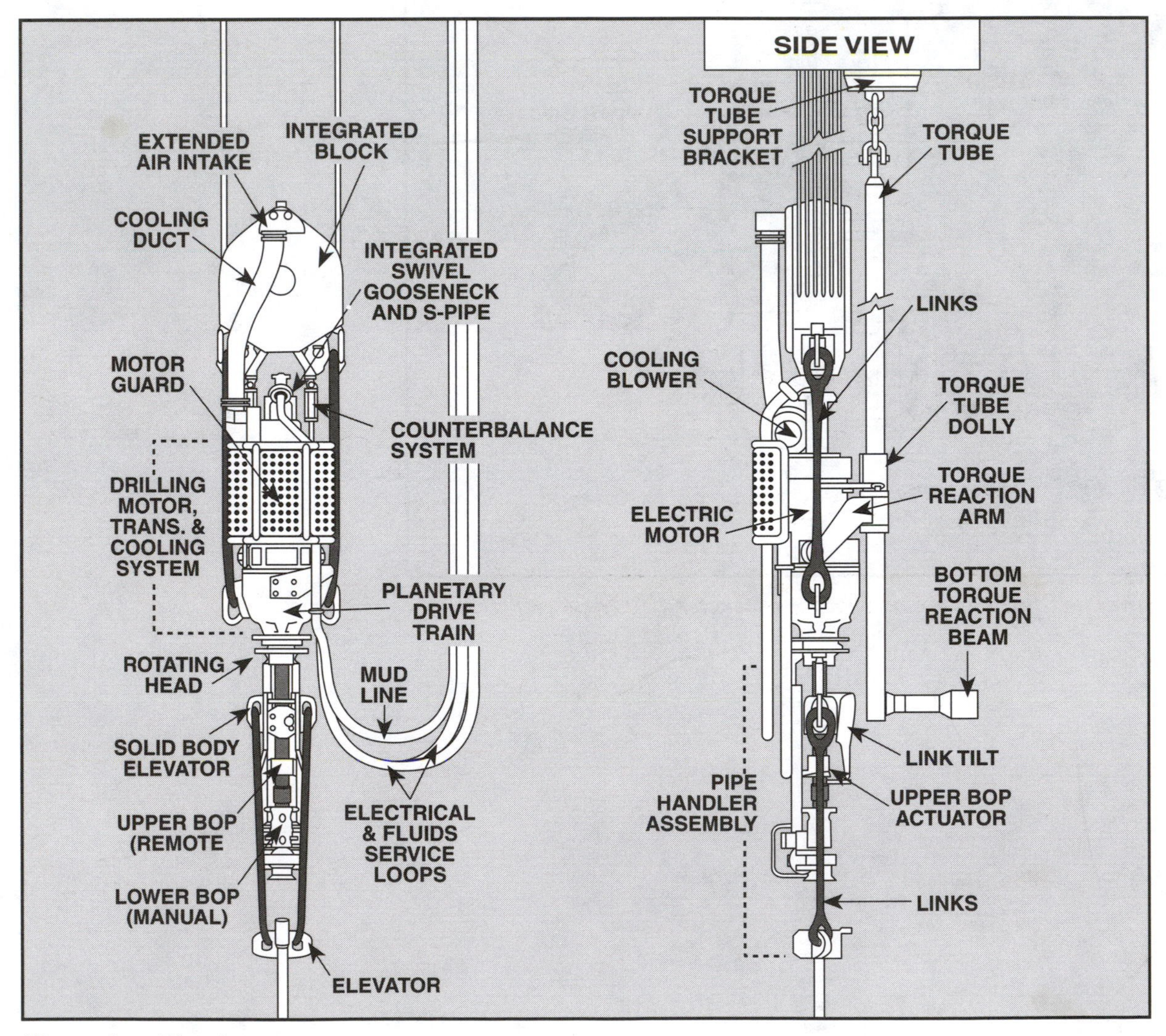

Figure 81. Top-drive parts

Several pieces of equipment are unique to the top drive. One is its built-in, high-torque motor (or motors), which powers a rotating head to which a drive shaft is connected. Electricity, hydraulic power, or mechanical power drives the motor or motors. The rotating head attaches to the swivel on one end and to the drill stem on the other through a drive shaft. Beveled gears inside the rotating head turn the bottom end of the drive shaft. The drive shaft rotates the drill pipe.

The top-drive guide rails and dolly are also distinctive. Rig owners mount the guide rails in the derrick. The top drive travels up and down in the guide rails. They stabilize the top drive and keep it from turning as it applies torque to turn the drill stem. The dolly is a frame that supports the heavy weight of the machinery as it moves vertically.

A special S-shaped pipe assembly connects the rotary hose to the top drive's swivel. In addition, the manufacturer installs a link-tilt device between the rotating head and the pipe-handling part of the top drive. When actuated, the link-tilt device positions the elevators for latching onto the pipe in the mousehole. The device is an air-powered arm that pushes the elevator links out over the pipe and then allows crew members to latch the elevators.

Another component that makes the top drive different is the pipe handler. It makes up and breaks out pipe from the rotating head. The pipe handler is also a torque wrench. The manufacturer also installs a counterbalance system. This system prevents damage to tool joint threads by softening the impact of the top drive as it meets the tool joints.

An important part of the top-drive system is the oil lubrication system, which cools the motor. Finally, an electrical system connects the top drive to the driller's controls.

Top drives are time and money savers in the long run, but they do have a few disadvantages. For example, top drives are expensive to buy and to maintain and, except for portable models, they are very large. Because standard top drives are large, rig owners cannot easily transport them from one rig to another. Also, installing a standard top drive requires that the derrick be of a certain minimum size. Except for portable top drives, even the smallest standard units are too large for many small rigs. What is more, because top drives are heavy, the rig's drilling line wears faster.

Because standard top drives are so large, contractors normally use them only on large rigs and deep wells. Installation frequently requires some modifications to the conventional drilling rig masts.

Manufacturers have developed a compact top drive that is portable. Because it is smaller than the other models and designed for easier installation and removal, contractors can purchase one and use it as needed on several rigs (fig. 82). This updated top drive is growing in popularity, but most models in use today are not the portable type.

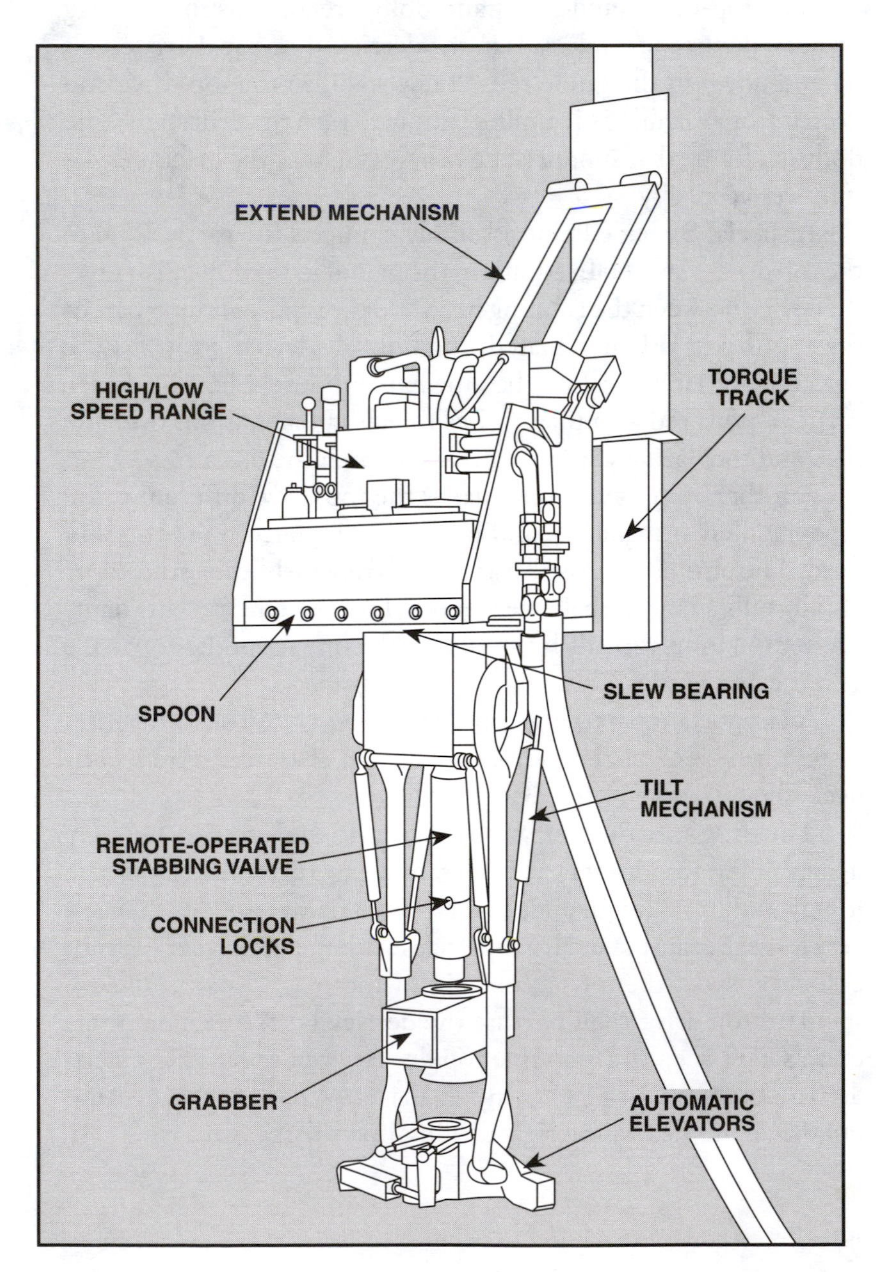

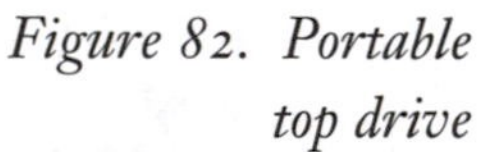
Figure 82. Portable top drive

How a Top Drive Works

Let's take a look at how a top drive makes up a three-joint stand of drill pipe (a thribble) during drilling. As a starting point, assume that the previously added stand has been drilled down until the top drive is just above the rig floor (fig. 83-1). Crew members made up the top drive's drive shaft to a saver sub, and they made up the saver sub to the last stand of pipe in the hole.

The driller stops rotating. The box of the pipe extends just above the rotary table. The rotary crew sets the slips to suspend the drill stem in the hole. Crew members engage the top drive's built-in pipe handler to break the drive shaft and saver sub out of the pipe's box. The rotating head spins out the connection.

With the top drive disconnected from the drill stem, the driller raises the top drive to the top of the thribble (fig. 83-2). (Remember that the crew previously made up this stand and set it back in the derrick). The derrickhand then stabs the drive shaft and saver sub into the stand. The top drive's built-in pipe handling equipment then spins up and torques the top drive's saver sub into the stand.

The driller raises the new stand with the top drive, and crew members position and stab the new stand into the pipe suspended in the rotary table's master bushing (83-3). The motor spins up the connection. The pipe handler's hydraulic tongs then torque up the connection. The driller raises the top drive a small distance to allow crew members to pull the slips. The driller actuates the rotating head and drive shaft to begin turning the drill stem, and drilling resumes (83-4, 83-5).

Figure 83. Making a connection with a top drive

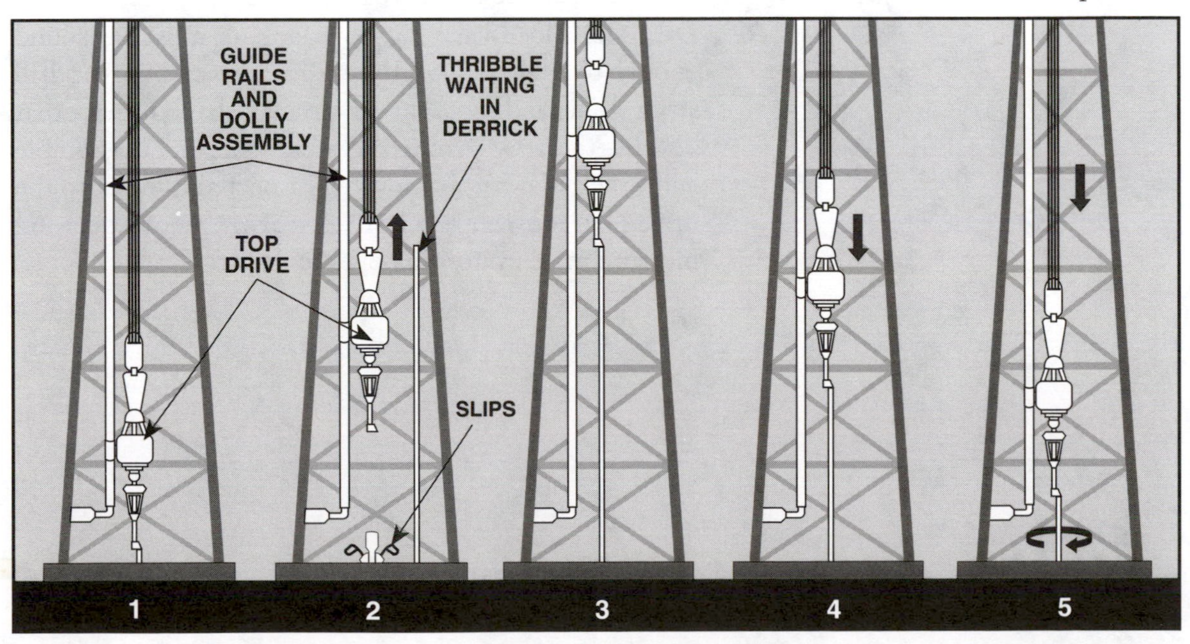

Maintenance

As with all rotating equipment, top-drive maintenance starts with the basics of regular lubrication and visual inspections. (See Appendixes H and I for lubrication, maintenance, and troubleshooting procedures for one brand of top drive.)

Visual Inspection

The electric drilling motor, air exhaust muffler, the rig air apparatus, the pipe handler, the inside blowout preventer, and the drill stem load-carrying components are some of the main areas that demand regular visual inspection.

1. Check the drilling motor's support bonnet and guide-dolly assembly daily for loose bolts and fittings.
2. The manufacturer installs air exhaust filters on the manifold to empty cooling air and motor exhaust. These filters are mufflers that reduce noise and contaminants. Periodically inspect, clean, and replace them as necessary.
3. Rig air powers the inside blowout preventer, the brake, and the torque wrench. Air valves, cylinders, and actuators for this pneumatic equipment require lubrication and clean air for operation.
4. Inspect the pipe handler daily for loose bolts and fittings. Make sure the hinge bolts are not worn or separating from the pipe handler.
5. Check the inside blowout preventer for broken connections; inspect the shoulders for gouges; and examine the outer surface for corrosion.
6. Drill stem load-carrying components must be sound enough to safely support the enormous weight of the drill string. The landing collar is a circular-shaped device that attaches the drive shaft to the rotating head. Disassemble and check it every six months. Inspect and measure the drive shaft for wear regularly. Also, regularly inspect and measure the elevator link eyes for wear.

Lubrication

A number of top-drive components need regularly scheduled lubrication. Among these are the motor, the washpipe, the rotating head, the link adapter, and the torque wrench. The elevator assembly, safety-valve actuator, inside blowout preventer valves, bail pins, and guide-dolly assembly also depend on regular lubrication.

The motor gear box is one of the most critical areas to check. Turn the motor off and check the oil daily to make sure the oil level is at the correct level indicated on the gauge. On a typical top drive, crew members should drain and refill the motor gearbox every 1,500 operating hours, or every three months. For more information on lubrication and maintenance, see Appendixes G and H.

To summarize—

Components of a typical top drive

- A drilling motor and transmission assembly
- Guide rails and dolly assembly
- An integrated swivel gooseneck and S-pipe
- A pipe handler
- A counterbalance system
- A motor cooling system
- A control system (electrical service loop and control panel)

Functions of the top drive

- Replaces the kelly
- Replaces the swivel
- Replaces the rotating function of the rotary table assembly

Conclusion

▼
▼
▼

The early days of rotary drilling overlapped the fading era of cable tool drilling. For a time, the rival technologies waged a battle across the oilfields of the United States. Rotary workers called cable tool crew members rope chokers, jar heads, and mail pouchers (after a brand of chewing tobacco). Cable tool workers branded rotary hands auger men, chain breakers, clutch stompers, twisters, and swivel necks. Feelings were so fierce in Electra, Texas, for example, that boardinghouses segregated the workers to keep fights from breaking out. One house was for swivel necks only; the other was reserved for mail pouchers.

Nearly a century later, the oilfield remains just as colorful, and drilling methods are still advancing. Rotary drilling eventually took its place as the industry standard. Now, top-drive technology is changing rotary drilling itself. The old debate over drilling gear continues as top-drive equipment improves and challenges the conventional rotary method. But the news media have not reported any instances of separate boardinghouses for top-drive hands and for rotary table hands—at least, not yet.

Appendix A

▼
▼
▼

Maintenance Checklist for One Brand of Rotary Table Assembly

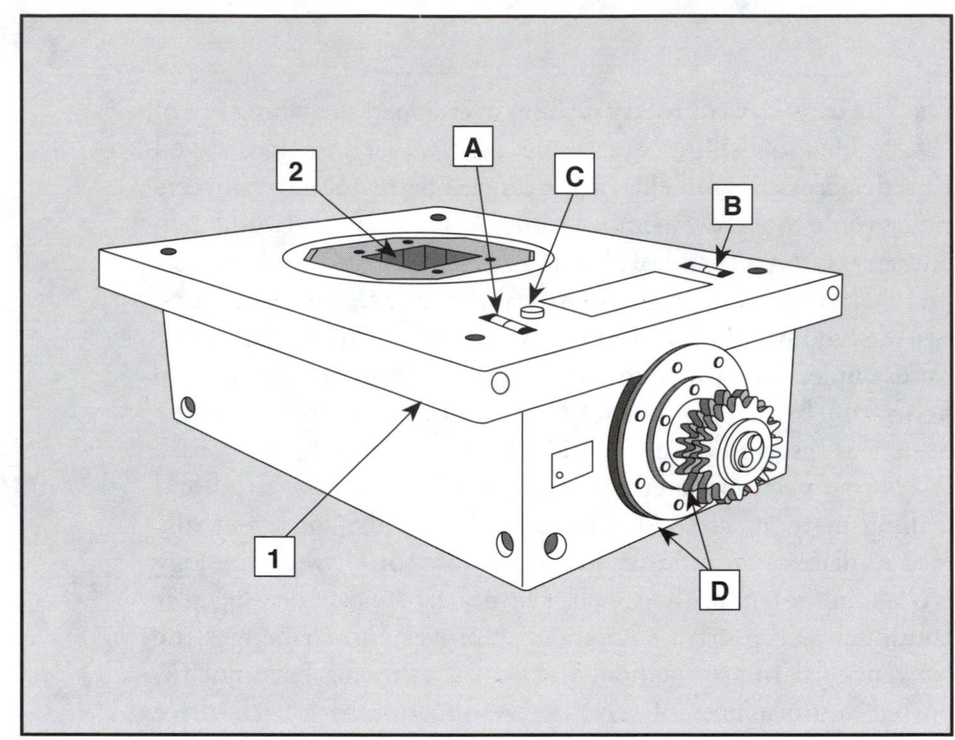

LUBRICATION SCHEDULE

Frequency	Check Point	Type	Procedure
Daily	A	Multipurpose grease*	Two fittings (left pawl)
	B	Multipurpose grease*	Two fittings (right pawl)
	C	AGMA EP gear lube**	Check fluid level while stationary.
Six months	D	AGMA EP gear lube***	Change oil: one drain plug is in the pinion housing and one plug is at the bottom of the main housing.

Surrounding Air Temperature Lubricant Selection

*Sodium-based multipurpose grease:	Below 32°F (0°C)—NLGI No. 1 Above 32°F (0°C)—NLGI No. 2
**Gear oil with rust inhibitor:	Below 60°F (16°C)—AGMA Mild EP-4 Above 50°F (10°C)—AGMA Mild EP-5

***Oil capacities: (National-Oilwell rotary tables)			
	C-175-S	5.0 gallons	(19 litres)
	C-175-L	5.5 gallons	(21 litres)
	C-205	7.5 gallons	(28 litres)
	C-275	8.0 gallons	(30 litres)
	C-375	7.0 gallons	(26.5 litres)
	D-375	8.0 gallons	(30 litres)
	C-495	17.0 gallons	(64 litres)
	D-495	19.0 gallons	(72 litres)

OPERATING MAINTENANCE
(refer to illustration on page 108)

Frequency	Check Point	Procedure
Monthly	1	Lift cover and remove excessive mud buildup around table rim and lock pawls.
	2	Inspect bore and master bushing for damage or excessive wear affecting fit.

PRECAUTIONARY MEASURES

Never operate a rotary table assembly that is contaminated with mud or water.
After a blowout or kick, check the sump for contamination.
Excessive oil consumption may indicate the need to replace the oil seal at the end of the pinion shaft housing.
Proper grounding procedures should be taken if welding on rotary tables to avoid arcing across bearings.
Do not engage lock pawls while the rotary is in motion.
Do not drive bushings into place with a sledge hammer.
Excessive wear between a split master bushing and the rotary table results in insufficient support for the bushing and the slips.

Maximum Recommended Static Loads and Zero-Load Operating Speeds

(National-Oilwell Rotary Tables)

C-175	250 tons (111 decanewtons)	500 rpm
C-205	350 tons (156 decanewtons)	400 rpm
C-275	500 tons (223 decanewtons)	350 rpm
C/D-375	650 tons (289 decanewtons)	350 rpm
C/D-495	800 tons (356 decanewtons)	300 rpm

(Courtesy National-Oilwell)

Appendix B

▼
▼
▼

Rotary Table Assembly Troubleshooting Guide

Common signs of trouble are listed below. If the problem persists, refer to the manufacturer's technical services department.

Observation	Probable Source	Remedy
Excessive oil consumption	Escaping from pressure system	Repair lubrication lines.
	Oil return passages plugged	Drain, flush, and refill unit.
	Damaged hold-down ring gasket	Replace gasket.
	Loose hold-down ring	Tighten and lock capscrews.
Contaminated oil	Lack of grease in upper labyrinth	Keep filled at all times.
	Damaged hold-down ring seal	Replace seal.
Locking or binding	Improper gear and pinion setting	Adjust as specified.
	Damaged pinion or ring gear	Replace gear set.
	Worn pinion shaft bearings	Replace bearings.
	Damaged thrust bearings	Completely replace bearings.
	Damaged locking device	Replace damaged parts.
Noisy operation	Improper gear and pinion setting	Adjust as specified.
	Damaged pinion or ring gear	Replace gear set.
	Worn pinion shaft bearings	Replace bearings.
	Damaged thrust bearings	Completely replace bearings.
	Loose pinion shaft sprocket	Tighten or replace sprocket.
	Loose hold-down ring	Tighten and lock capscrews.
	Echoing drive noises	Check chain, coupling, etc.

Rotary Table Assembly Troubleshooting Guide, cont.

Observation	Probable Source	Remedy
Excessive heat	Insufficient lubrication	Maintain proper oil level.
		Clean suction strainer.
		Replace filter cartridge.
		Replace lubrication lines.
		Replace lubricating oil pump.
		Clean magnetic filter.
		Adjust pump relief valve.
	Insufficient pinion bearing grease	Maintain proper oil level.
	Excessive pinion bearing grease	Plugged relief line or fitting

(Courtesy National-Oilwell)

Appendix C

▼
▼
▼

Lubrication and Maintenance Checklist for One Brand of Swivel

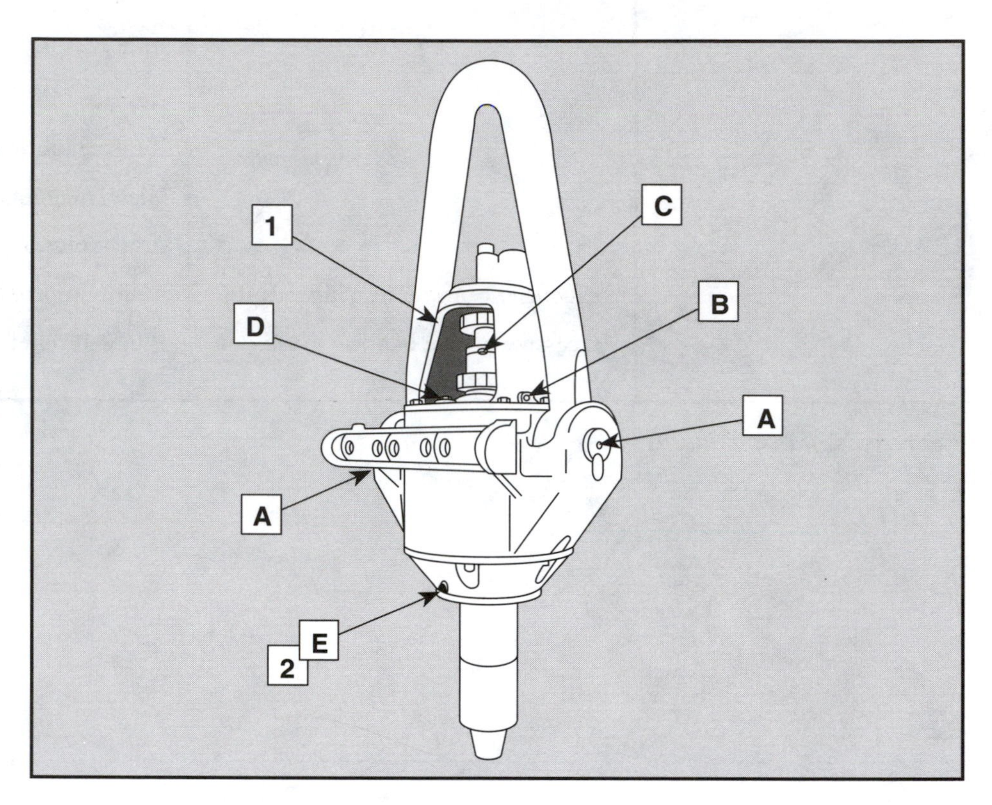

LUBRICATION SCHEDULE

Frequency	Check Point	Type	Procedure
Daily	A	Multipurpose lithium-based grease*	1 to 3 strokes of a hand grease gun (bail pins)
	B	Multipurpose lithium-based grease*	1 to 3 strokes of a hand grease gun (upper oil seals)
	C	Multipurpose lithium-based grease*	1 to 3 strokes of a hand grease gun (packing assembly)
	D	AGMA EP gear lubricant**	Check fluid level at rest with gauge screwed in hand-tight.
1,000 hours	E	AGMA EP gear lubricant***	Drain oil, flush, and refill with fresh oil.

Surrounding Air Temperature Lubricant Selection

*Multipurpose lithium-based grease:	Below 32°F (0°C)—NLGI No. 1 Above 32°F (0°C)—NLGI No. 2
**Gear oil with rust inhibitors:	Below 60°F (16°C)—AGMA Mild EP-4 Above 50°F (10°C)—AGMA Mild EP-5
***Oil capacities (National-Oilwell swivels):	P-200—10 gallons (38 litres) P-300—14 gallons (53 litres) P-400—16 gallons (61 litres) P-500—24 gallons (91 litres) P-650—30 gallons (114 litres)

SWIVEL OPERATING MAINTENANCE

(refer to illustration on page 112)

Frequency	Check Point	Procedure
Monthly	1	Check pressure-relief fitting (located on opposite side from lube fitting) to be sure it is not damaged or plugged.
1,000 hours	2	Check bar magnets on drain plug (metal particles on magnets could indicate excessive bearing wear).

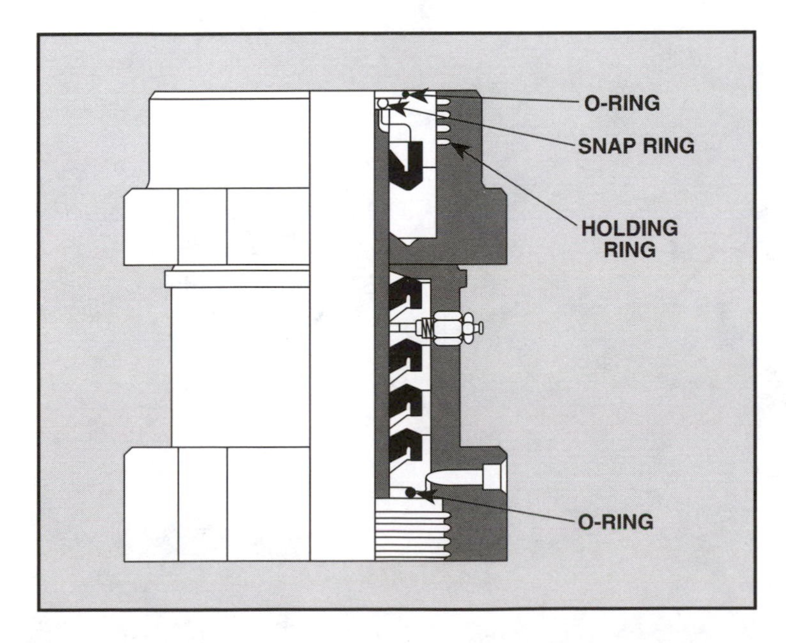

Replacing Washpipe Packing

When it is necessary to replace the washpipe packing, check the assembly to be sure that the new O-ring is installed on the top and the bottom of the packing assembly and that the finished arrangement looks like the figure at left. Inspect all parts for excessive wear, corrosion, or washout. Pay particular attention to the holding-ring splines and snap groove. Hand-pack grease into cavities formed by each packing lip and spacer before placing into service.

PRECAUTIONARY MEASURES

Check the swivel sub connection for metal-to-metal (shoulder-to-shoulder) tightness to avoid thread damage or failure.

If flow from the oil drain hole is restricted, contaminants have likely accumulated that could shorten bearing life. Disassemble the swivel and thoroughly clean and inspect it.

Grease bail pins with the bail in an at-rest position. With the bail at rest, the grease can flow into heavy load–carrying areas.

Check the main bearings for wear after jarring operations.

If welding on the swivel, be sure to properly ground the operation so that no arcs flash across the bearings.

Always store the swivel in an upright position with a full fresh change of oil to prevent bearing corrosion.

(Courtesy National-Oilwell)

Appendix D

Swivel Troubleshooting Guide

Trouble	Possible Causes	Remedy
Premature failure of washpipe packing	Lack of lubrication every tour	Grease at least once.
	Insufficient tightening of packing housing, which allows packing to rotate in cartridge	Tighten packing housing sufficiently to prevent the packing from rotating.
	Excessive tightening of packing housing, which causes damage to dowels in packing rings; packing becomes overcompressed	Replace rings and packing if necessary. Tighten cartridge moderately. Make sure it is tight enough to prevent packing rotation.
	Worn washpipe	Replace washpipe; also install new packing.
	Misalignment of washpipe from distorted housing cap, which can be caused by rough handling or accident; the condition is indicated by unequal circumferential wear on the washpipe.	Immediately repair or replace housing cap and washpipe.
	Improper bearing adjustment; a vertical clearance in excess of 0.007 inches (0.18 millimetres) is recommended.	Check with manufacturer for procedure for bearing adjustment.
Oil leaks at oil seals	Worn or damaged oil seals	Replace oil seals. Take care not to damage sealing lips.
	Absence of breather, or breather is plugged; the heat generated in rotating causes the lubricant to expand and build up pressure.	Install the breather supplied with the swivel or clean the breather by washing it thoroughly with kerosene.

(Courtesy National-Oilwell)

Appendix E

▼
▼
▼

Hydraulic Spinning Wrench Lubrication

To ensure trouble-free operation, lubricate the spinning wrench as indicated in the schedule. Use high-quality multipurpose water-resistant grease. Apply the grease to the lubrication points until fresh grease appears near the lubrication point.

LUBRICATION SCHEDULE

Item No. and Name	Number of Lube Points	Application	Lube Cycle
1. Pressure rollers	2	Multipurpose water-resistant grease	Every trip
2. Pressure roller pivots	2	Multipurpose water-resistant grease	Every trip
3. Drive roller pivots	2	Multipurpose water-resistant grease	Every trip
4. Clamp cylinder link pivots	2	Multipurpose water-resistant grease	Every trip
5. Clamp beam pivots	2	Multipurpose water-resistant grease	Every trip
6. Gear case	1	Drain and refill (3 quarts SAE 90)	Every three months
7. Hose connection swivel fittings	2	Multipurpose water-resistant grease	Every trip

(Courtesy Varco BJ)

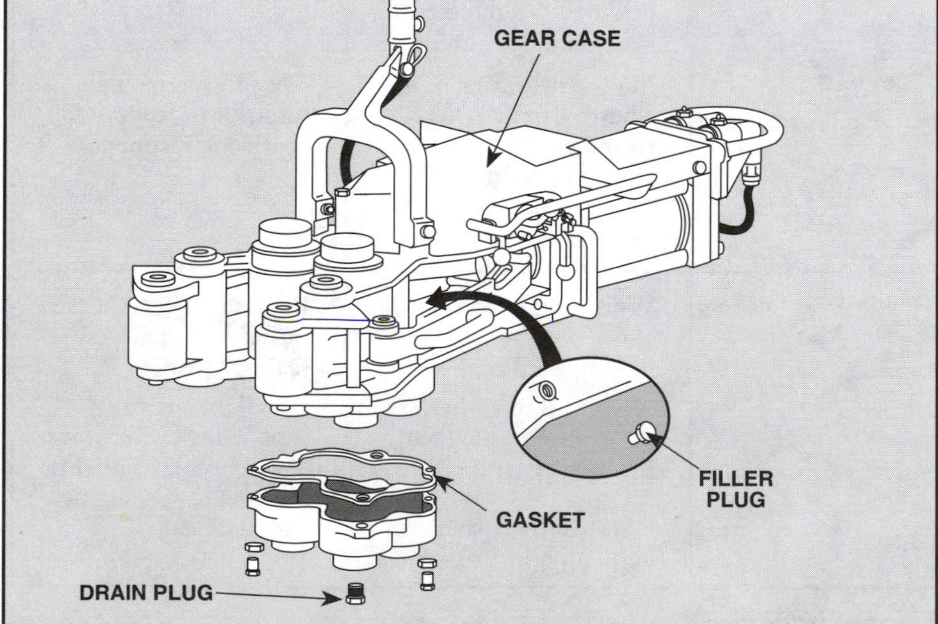

Hydraulic Spinning Wrench Maintenance

Every three months:

1. Drain the oil from the gear case and remove the gear case cover.

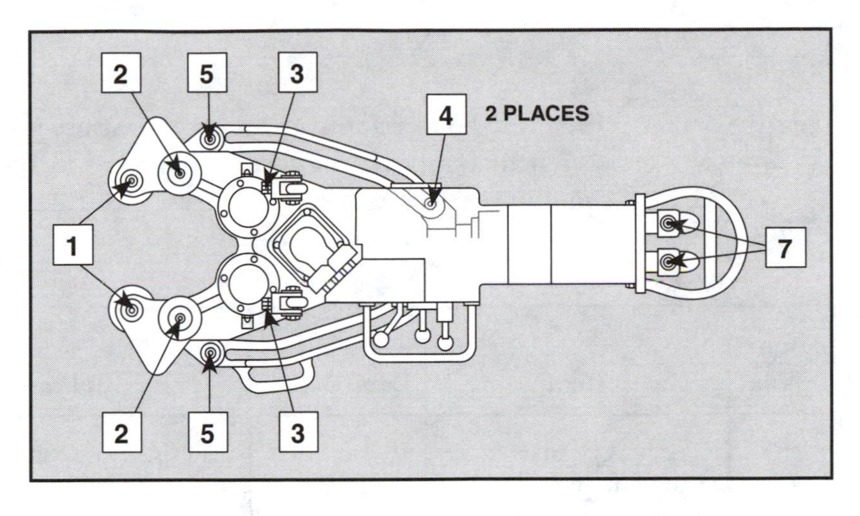

2. Check the torque of the motor mounting bolts for 140 to 150 foot-pounds (190 to 203 newton metres).
3. Check the gear train for any signs of unusual wear. Note the condition of the intermediate shafts and bearings. If any unusual wear is noted in the gear case, determine the cause and correct it.
4. Reinstall the gear case cover with a new gasket.
5. Fill the gear case with 3 quarts (2.8 litres) of SAE 90 oil.
6. Check the hydraulic power supply using the instructions in the manufacturer's service manual.

TROUBLESHOOTING

Symptom	Cause	Remedy
Arms do not clamp or unclamp; or motor does not run forward, reverse, or both.	Flow restricted	Locate cause of restriction and correct it.
	Power supply not operating properly	Check power supply operation.
	Valve linkage defective	Repair or replace linkage.
Arms or motor creep with valve in neutral position.	Detent capscrews loose in valve, causing spool misalignment	Tighten detent capscrews and safety wire.

(Courtesy Varco BJ)

Appendix F

▼
▼
▼

Torque Wrench Lubrication

Thorough lubrication is necessary to ensure proper operation of the torque wrench. Lubricate the tool as indicated in the table. Use a high-quality multipurpose water-resistant grease, NGLI grade 2.

Item No.	Item	No. of Lube Points	Application	Lube Cycle
1	Torque cylinders (2 on trunnion mounting, 1 on rod-end pin)	6	Multipurpose water-resistant grease	Every trip
2	Gates (2 on latch, 1 on gate hinge)	6	Multipurpose water-resistant grease	Every trip
3	Vertical positioning (1 on each pulley shaft)	6	Multipurpose water-resistant grease	Every trip
4	Body-to-body contact surfaces (apply with upper body fully rotated)	as required	Multipurpose water-resistant grease	Every trip
5	Pressure-reducing valve (PRV)	1	SAE 10 oil	Every trip
6	Valve handle links and spools	as required	SAE 10 oil	Every trip

(Courtesy Varco BJ)

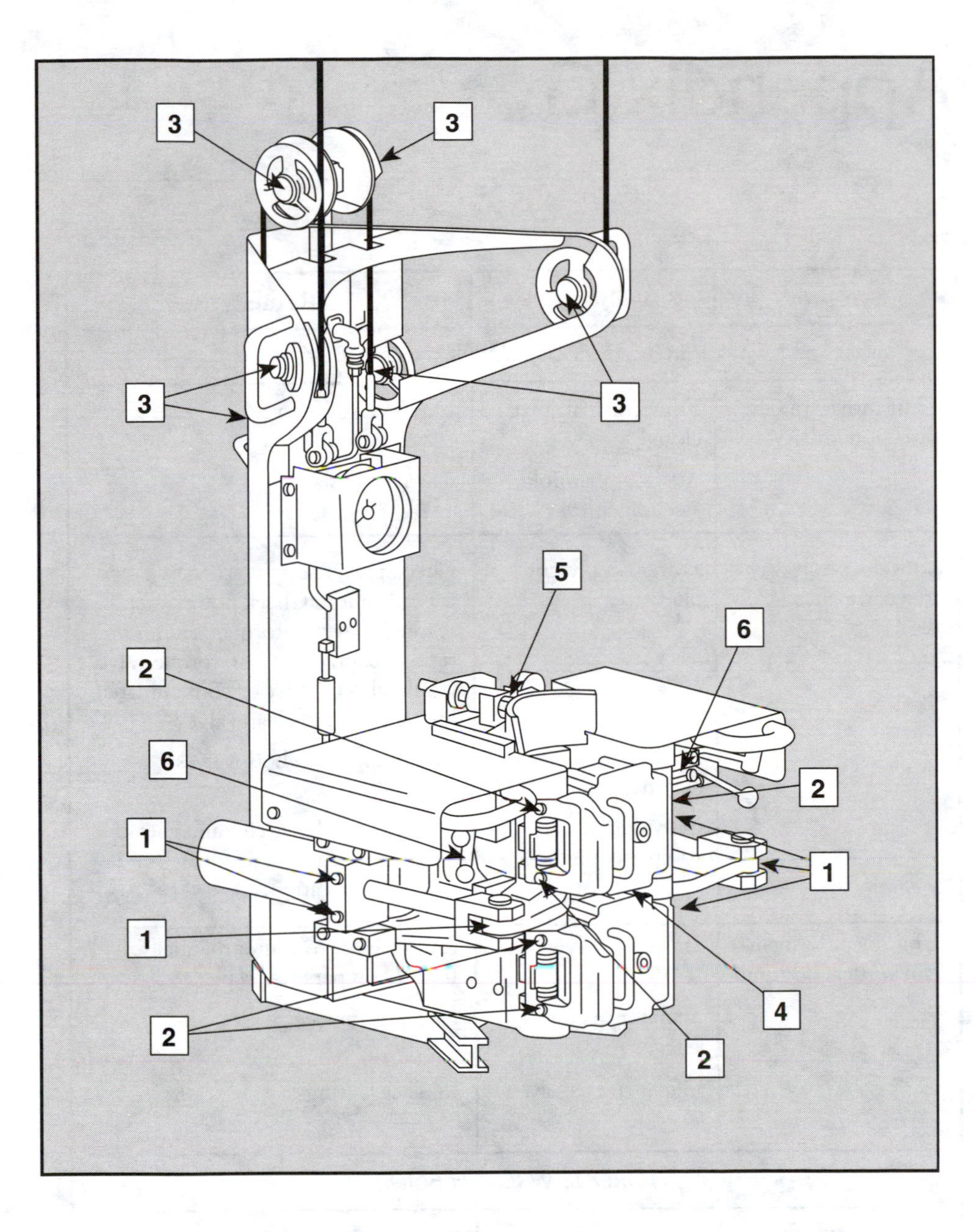
3
3
3
3
3
5
6
2
6
2
1
1
1
1
4
2
2

Appendix G

Torque Wrench Troubleshooting Guide

Symptom	Probable Cause	Remedy
VERTICAL POSITIONING ASSEMBLY		
Unit moves in one direction only.	Flow control valve closed	Open valve if closed.
	Actuator manifold assembly dirty	Clean or replace.
Unit does not move in either direction.	Power supply not operating	Check power supply pressure gauge for indication. Check power supply manual. Operate torque lever; if no response, problem is in torque valve. Could also be dirty or contaminated line. Check and clean valve.
	Both control valves closed	Check and open both valves.
	Individual component failure	Trace hydraulic circuit and check all components for adequate fluid flow and operation.
Unit does not maintain vertical position.	Sticky lift valve	Return valve to center manually. If problem persists, replace valve.
	Actuator manifold leaky	Clean or replace.
	Lift valve excessively leaky	Replace O-rings.
TORQUE WRENCH BODIES		
Torque cylinders will not make up.	Pressure-reducing valve (PRV) setting incorrect	Check that PRV setting is not too low.
	Dump valve stuck open (pressure does not build up)	Manually actuate valve; replace if faulty.
	PRV stuck closed, is dirty or defective	Check that PRV is screwed in; clean or replace valve.
Torque cylinders will not break out.	PRV stuck closed, is dirty or defective	Check that PRV is screwed in; clean or replace valve.

Torque Wrench Troubleshooting Guide, cont.

Symptom	Probable Cause	Remedy
TORQUE WRENCH BODIES, cont.		
Torque cylinders drift in makeup direction.	Sticky torque valve	Return valve to center position manually; if problem persists, replace it.
Torque cylinders drift in breakout direction.	Sticky torque valve	Return valve to center position manually; if problem persists, replace it.
	Excessive back-pressure on return line	Check hydraulic power supply for defect or malfunction.
Torque gauge does not indicate pressure during makeup.	Dirty inlet port to gauge	Clean inlet port or replace gauge.
	Torque cylinders reached end of stroke and actuated dump valve	Reset for additional stroke.
	Dirty orifice	Clean.
	Gauge damper closed	Open damper (rotate counter-clockwise).
Torque gauge does not return to zero.	Gauge not adjusted to zero	Rotate zero-adjust knob on back of gauge case. Recheck torque setting.
	Gauge defective	Replace it.
Torque gauge fluctuates.	Gauge damper out of adjustment	Adjust by pushing in to engage threads and rotate clockwise to smooth fluctuations.
Clamp valve doesn't shift or overtravels and will not return.	Defective detent in control valve	Replace valve.
Joint is not centered in wrench when clamped.	Centering-button spring broken or has taken a permanent set	Repair or replace.
Hydraulic fluid leakage around either end of clamp cylinders	Worn or defective seals	Replace seals.
Jaws do not grip joint.	Worn or broken dies	Replace dies.
	Incorrect spacer arrangement	Check die and spacer arrangement.

(Courtesy Varco BJ)

Appendix H

▼
▼
▼

Top-Drive Maintenance and Lubrication

Perform the following safety checks and maintenance procedures on a regular basis.

1. *DC Drilling Motor*: Thoroughly inspect the TDS, the motor-support bonnet, and the guide-dolly assembly for loose bolts and fittings daily. Replace any safety wire or cotter pins removed for repairs.
2. *Air Exhaust Muffler*: Periodically remove the air exhaust mufflers and replace them.
3. *Derrick Termination Kit, Air Filter-Regulator-Lubricators*: Inspect monthly for damage to the filter bowl, overall body, or inlet and outlet fittings. Replace any damaged, corroded, or improperly functioning components. Adjust the pressure regulator outlet pressure to 100 to 120 pounds per square inch (690 to 825 kilopascals) monthly. Clean and drain the filter weekly. Fill the lubricator with petroleum-based hydraulic oil with a fluid viscosity of 100 to 200 Saybolt seconds universal (SSU) at 100°F (38°C) (ISO 32/34).
4. *Pipe Handler*: Thoroughly inspect for loose bolts and fittings daily. Replace any safety wire or cotter pins removed during repairs. Make sure hinge pins (if any are present) are not loose because of either excessive wear in the bore of the clamp clevis or a broken retaining bolt. WARNING: *Inspect the link tilt intermediate stop and adjustment device for overall integrity weekly. Replace the components if the threaded rod is excessively loose in the threaded hole of the pivot arm. Failure to perform this inspection and component replacement can result in injury to rig personnel.*
5. *Landing Collar*: Disassemble and inspect the landing collar every six months. After disassembly, inspect all landing collar parts for wear, damage, or corrosion. Replace the appropriate parts as necessary and regrease prior to assembly. Bend all 32 tabs.
6. *Safety Valve (IBOP) Inspection Procedures*: Whenever connections are broken, clean and check for—
   a. galling on the threads
   b. stretching or other abnormal conditions
   c. marks, gouges, or other damage on the shoulders
   d. excessive tong marks and corrosion on the outer surface
   e. wear on the splines of the upper safety valve
7. *Cooling System*: On rigs with closed-loop cooling systems, check the heat exchangers for water leakage weekly and remove the access covers and operate the blower to remove carbon dust buildup every 500 operating hours.

(Courtesy Varco BJ)

LUBRICATION SCHEDULE

Item	Frequency	Location	No. of Pts.
1	Daily	Torque wrench	12
2	Daily	Gearbox oil (check level)	1
3	Daily	Rotating head	7
4	Daily	Safety valve actuator cranks	2
5	Daily	Hydraulic pressure filter†	–
	Daily	Wash pipe assembly*	1
6	Each trip	BNC drill pipe elevator	7
2	Weekly	Gearbox (shift gears)	–
7	Weekly	Dolly motor trunnions	2
8	Weekly	Air filter-regulator-lubricator†*	–
9	Weekly	IBOP actuator arms	4
10	Weekly	Link adapter	2
	Weekly	Master bushing wear guide*	4
11	Weekly	Link tilt (check condition)	–
12	Weekly	Guide dolly rollers	20
13	Weekly	Bail pins	2
	750 hrs or 90 days	Drilling motor pinion bearing*‡	–
	1,500 hrs or 6 mos	Drilling motor commutator bearing*‡	–
14	Monthly	Motor frame and dolly hinges	4
15	Monthly	Cooling system cleaning	–
16	3 months	AC blower motor	2
2	3 months	Change gear box oil	1
	3 months	Gear oil filter and suction strainer†*	1

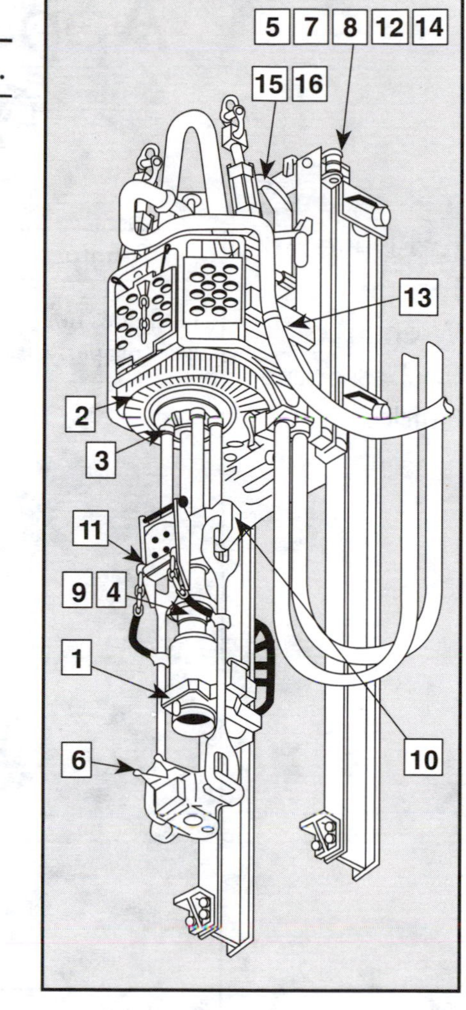

†Check condition and replace as necessary
*Item not shown in illustration
‡Use 2 ounces by weight Shell Cyprina RA™

(Courtesy Varco BJ)

Appendix I

Top-Drive Torque Wrench Troubleshooting Guide

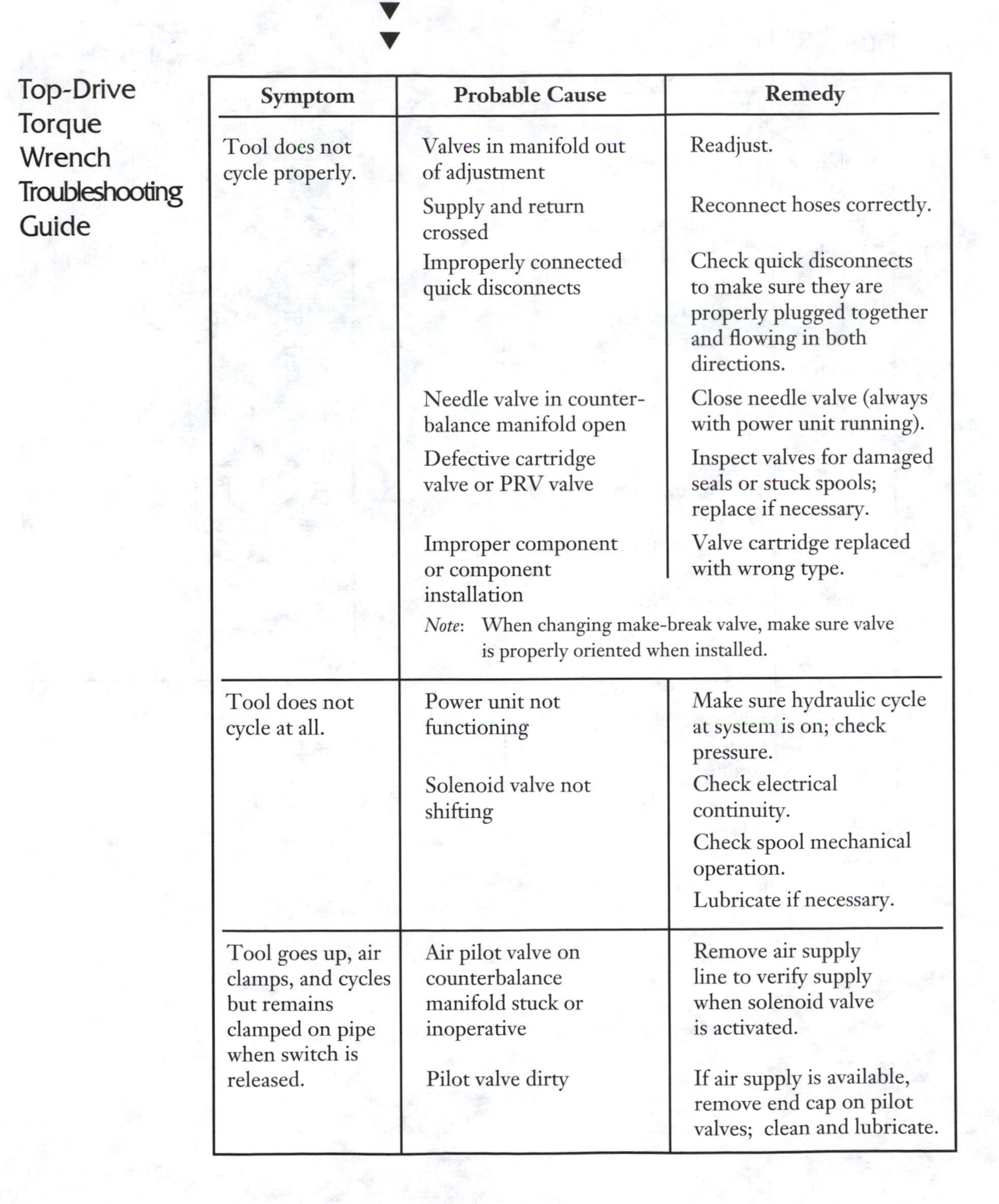

Symptom	Probable Cause	Remedy
Tool does not cycle properly.	Valves in manifold out of adjustment	Readjust.
	Supply and return crossed	Reconnect hoses correctly.
	Improperly connected quick disconnects	Check quick disconnects to make sure they are properly plugged together and flowing in both directions.
	Needle valve in counter-balance manifold open	Close needle valve (always with power unit running).
	Defective cartridge valve or PRV valve	Inspect valves for damaged seals or stuck spools; replace if necessary.
	Improper component or component installation	Valve cartridge replaced with wrong type.
	Note: When changing make-break valve, make sure valve is properly oriented when installed.	
Tool does not cycle at all.	Power unit not functioning	Make sure hydraulic cycle at system is on; check pressure.
	Solenoid valve not shifting	Check electrical continuity.
		Check spool mechanical operation.
		Lubricate if necessary.
Tool goes up, air clamps, and cycles but remains clamped on pipe when switch is released.	Air pilot valve on counterbalance manifold stuck or inoperative	Remove air supply line to verify supply when solenoid valve is activated.
	Pilot valve dirty	If air supply is available, remove end cap on pilot valves; clean and lubricate.

Top-Drive Torque Wrench Troubleshooting Guide, cont.

Symptom	Probable Cause	Remedy
Tool goes up, air clamps, and cycles but remains clamped on pipe when switch is released (cont.)	Improperly connected quick disconnects	If air pilot valve is operable, hook both pipe handler hoses together and verify flow in both directions by activating switch with hydraulics on. If flow occurs in only one direction, check quick disconnects.
	Hydraulic valve not shifting	Repair or replace.
Saver sub breaks out instead of tool joint.	Saver sub not made up properly	Make up saver sub according to the procedure described in the manufacturer's service manual; or increase previous makeup torque by 10% to maximum torque recommended in the service manual.
Tong dies break or score drill pipe.	Valves in manifold out of adjustment	Readjust according to procedure given in service manual.
	Tong dies worn	Replace tong dies.
	Connection overtorqued	If saver sub is made up to recommended torque and continues to break out instead of drill pipe connection, use manual tongs until manufacturer's service representative can make repair.
Die retainer and die retainer bolts damaged while breaking out connections.	Improper saver sub length or machining	Saver sub length too long; sub should also have proper chamfer, or beveled edge, on both shoulders.
Tong wrench does not lift.	Oil bypass in pipe handler hydraulic components	Listen for oil bypassing in pipe handler cylinders or manifold; locate and repair bypass as necessary.

(Courtesy Varco BJ)

Glossary

▼
▼
▼

A

American Petroleum Institute (API) *n*: founded in 1920, this national oil trade organization is the leading standardizing organization on oilfield drilling and producing equipment.

API *abbr*: American Petroleum Institute.

B

backup tongs *n pl*: the tongs latched on the drill pipe joint hanging in the rotary by the slips, used to keep the pipe from turning as the makeup or breakout tongs (the lead tongs) apply torque to make up or break out the tightened tool joint connection. Compare *lead tongs*.

bail *n*: a cylindrical steel bar (similar to the handle or bail of a bucket, only much larger) that supports the swivel and connects it to the hook. Sometimes, the two cylindrical bars that support the elevators and attach them to the hook are called bails or links. *v*: to recover bottomhole fluids, samples, or drill cuttings by lowering a cylindrical vessel called a bailer to the bottom of a well, filling it, and retrieving it.

bail pin *n*: a large steel dowel (pin) that attaches the swivel's bail to the swivel's body. Normally, swivels have two bail pins, one for each side of the bail where it attaches to the body. See *bail*.

bail throat *n*: the inside curve in the swivel's bail where the bail hangs from the traveling block's hook. See *bail*.

ball bearing *n*: a bearing in which a finely machined shaft (a journal) turns on freely rotating hardened-steel spheres that roll easily within a groove or track (a race) and thus convert sliding friction into rolling friction.

bevel gear *n*: one of a pair of toothed wheels whose working surfaces are inclined to nonparallel axes.

bit *n*: the cutting or boring element used in drilling oil and gas wells. The bit consists of the cutting element and the circulating element. The circulating element permits the passage of drilling fluid and utilizes the hydraulic force of the fluid stream to improve drilling rates. In rotary drilling, several drill collars are joined to the bottom end of the drill pipe column. The bit is attached to the end of the drill collar.

bit breaker *n*: a special device that fits into a bit breaker adapter (a plate that goes into the rotary table) and conforms to the shape of the bit. Rig workers place the bit to be made up or broken out of the drill stem into the bit breaker and lock the rotary table to hold the bit breaker and bit stationary so that they can tighten or loosen the bit.

bit breaker adapter *n*: a heavy plate that fits into the rotary table and holds the bit breaker, a device used to hold the bit while it is being made up or broken out of the drill stem.

bit nozzle *n*: see *nozzle*

block *n*: any assembly of pulleys on a common framework; in mechanics, one or more pulleys, or sheaves, mounted to rotate on a common axis. The crown block is an assembly of sheaves mounted on beams at the top of the derrick. The drilling line is reeved over the sheaves of the crown block alternately with the sheaves of the traveling block, which is hoisted and lowered in the derrick by the drilling line. When elevators are attached to a hook on a conventional traveling block, and when drill pipe is latched in the elevators, the pipe can be raised or lowered in the derrick or mast.

blowout *n*: an uncontrolled flow of gas, oil, or other well fluids into the atmosphere or into an underground formation the wellbore has penetrated. A blowout, or gusher, occurs when formation pressure exceeds the pressure applied to it by the column of drilling fluid. A kick warns of an impending blowout. See *kick*.

bottleneck *n*: an area of reduced diameter in pipe, brought about by excessive longitudinal strain or by a combination of longitudinal strain and the swaging action of a body. A bottleneck may result if the downward motion of the drill pipe is stopped with the slips instead of the brake.

box *n*: the female section of a tool joint. See *tool joint*.

break out *v*: to unscrew one section of pipe from another section, especially drill pipe while it is being withdrawn from the wellbore. During this operation, the tongs are used to start the unscrewing operation.

breakout tongs *n pl*: tongs that are used to start unscrewing one section of pipe from another section, especially drill pipe coming out of the hole. Compare *makeup tongs*. See also *tongs*.

buck up *v*: to tighten up a threaded connection (such as two joints of drill pipe).

C

cable-tool drilling *n*: a drilling method in which the hole is drilled by the rig's equipment dropping a sharply pointed and heavily weighted bit on bottom. The bit and weight are attached to a cable, and the rig's equipment repeatedly drops the cable to drill the hole.

casing *n*: steel pipe placed in an oil or gas well as drilling progresses to prevent the wall of the hole from caving in during drilling and to provide a means of extracting petroleum if the well is productive.

cathead *n*: a spool-shaped attachment on a winch or drawworks around which rope for hoisting and pulling is wound.

catline *n*: a hoisting or pulling line powered by the cathead and used to lift heavy equipment on the rig. See *cathead*.

circulating fluid *n*: see *drilling fluid; mud*.

come out of the hole *v*: to pull the drill stem out of the wellbore. This withdrawal is necessary to change the bit, change from a core barrel to the bit, run electric logs, prepare for a drill stem test, run casing, and so on.

conventional rotating system *n*: the rotary drilling system that uses a conventional swivel, a kelly, a kelly drive bushing, a master bushing, and a rotary table to turn the drill stem and bit. Compare *top drive*.

counterbalance *n*: see *tong counterbalance*.

cutters *n pl*: on a bit used on a rotary rig, the elements on the end (and sometimes the sides) of the bit that scrape, gouge, or otherwise remove the formation to make hole.

cuttings *n pl*: chips of rock made by a drill bit as the bit drills ahead and brought to the surface by the drilling mud.

D

detent *n*: a mechanism that keeps one part in a certain position relative to that of another; it can be released by applying force to one of the parts.

dies *n pl*: see *insert*.

dolly assembly *n*: on a top-drive unit, the device that attaches the unit to the guide rails. See *guide rails, top drive*.

double-plane roller assembly *n*: in a kelly bushing, a double set of roller assemblies that are installed on two levels, or planes, within the bushing. Compare *single-plane roller assembly*. See also *kelly bushing, roller assembly*.

drawworks *n*: the hoisting mechanism on a drilling rig. It is essentially a large winch that spools off or takes in the drilling line and thus raises or lowers the drill stem and bit.

drill ahead *v*: to continue drilling operations.

drill collar *n*: a heavy, thick-walled tube, usually steel, used between the drill pipe and the bit in the drill stem to provide weight to the bit.

drill collar slips *n pl*: see *slips*.

driller *n*: the employee directly in charge of a drilling or workover rig and crew. Main duty is operation of the drilling and hoisting equipment, but also responsible for the downhole condition of the well, operation of downhole tools, and pipe measurements.

drilling fluid *n*: circulating fluid, one function of which is to force cuttings out of the wellbore and to the surface. It also serves to cool the bit and counteract downhole formation pressures. While a mixture of clay, water, and other chemical additives is the most common drilling fluid, wells can also be drilled using air, gas, or oil as the drilling fluid. Also called circulating fluid, drilling mud. See *mud*.

drill pipe *n*: the heavy seamless tubing used to rotate the bit and circulate the drilling fluid. Joints of pipe about 30 feet long are coupled together by means of tool joints.

drill pipe safety valve *n*: see *lower kelly cock*.

drill stem *n*: all members in the assembly used for drilling by the rotary method from the swivel to the bit, including the kelly, drill pipe and tool joints, drill collars, stabilizers, and various subsequent items. Compare *drill string*.

drill string *n*: the column, or string, of drill pipe with attached tool joints that transmits fluid and rotational power from the kelly to the drill collars and bit. Often, especially in the oil patch, the term is loosely applied to include both drill pipe and drill collars. Compare *drill stem*.

drive bushing *n*: see *kelly bushing*.

drive shaft *n*: a shaft that transmits mechanical power.

E

elevators *n pl*: on conventional rotary rigs and top-drive rigs, hinged steel devices with manual operating handles that crew members latch onto a tool joint (or a sub). Since the elevators are directly connected to the traveling block, or to the integrated traveling block in the top drive, when the driller raises or lowers the block or the top-drive unit, the drill pipe is also raised or lowered.

extra-long rotary slips *n pl*: slips for drill pipe that fit into the tapered insert bowl of a four-pin master bushing and whose taper length is 12¾ inches (324 millimetres). Compare *long rotary slips*. See also *master bushing, slips*.

F

floor crew *n*: those workers on a drilling or workover rig who work primarily on the rig floor. See *rotary helper*.

four-pin kelly bushing *n*: a kelly bushing that has four steel dowels, or pins, that fit into corresponding holes in the master bushing. When the pins are engaged with the holes, the rotating master bushing also turns the kelly bushing, which then turns the kelly and the drill stem. See *kelly, kelly bushing, master bushing*.

four-pin master bushing *n*: a master bushing that has four holes symmetrically positioned on its outside perimeter and into which fit four corresponding steel dowels, or pins, on the kelly bushing. When the pins are engaged into the holes and the master bushing turns, the kelly bushing also turns. See *kelly bushing, master bushing*.

G

go in the hole *v*: to lower the drill stem into the wellbore.

gooseneck *n*: an erosion-resistant nozzle on the swivel that conducts high-pressure, high-volume drilling mud into the swivel.

grease fitting *n*: a device on a machine designed to accept the hose of a grease gun so that grease can be added to the part in need of lubrication.

guide line *n*: on a spinning chain, a piece of fiber rope (soft line) that is attached to the end of the chain that is wrapped around the tool joint. When the makeup cathead pulls the chain off the drill pipe, a rotary helper uses the guide line to control the chain's movement.

guide rails *n pl*: on some top drives, the steel tracks on which the top-drive unit travels up and down when the driller raises or lowers the traveling block. The rails also keep the unit from turning when its motor rotates the drill stem. Compare *torque track, torque tube*. See also *top drive*.

H

hexagonal kelly *n*: a kelly with a six-sided (hexagonal) cross section. Compare *square kelly*.

hinged master bushing *n*: a two-piece master bushing that has a jointed, swinging device (a hinge) on each half into which large pins fit to hold the bushing together. A two-piece insert bowl to hold the slips fits inside this type of master bushing. Compare *solid master bushing, split master bushing*. See also *insert bowl, master bushing, slips*.

hoist *n*: 1. an arrangement of pulleys and wire rope or chain used for lifting heavy objects; a winch or similar device. 2. the drawworks. See *drawworks*.

hook *n*: a large, hook-shaped device from which the swivel is suspended. It is designed to carry maximum loads ranging from 100 to 650 tons (91 to 590 tonnes) and turns on bearings in its supporting housing. A strong spring within the assembly cushions the weight of a stand (90 feet or 27 metres) of drill pipe, thus permitting the pipe to be made up and broken out with less damage to the tool joint threads. Smaller hooks without the spring are used for handling tubing and sucker rods. See also *stand* and *swivel*.

housing *n*: something that covers or protects, such as the casing for a mechanical part.

hydraulic *adj*: 1. of or relating to water or other liquid in motion. 2. operated, moved, or effected by water or liquid.

hydraulic torque wrench *n*: a hydraulically powered device that can break out or make up tool joints and assure accurate torque. It is fitted with a repeater gauge so that the driller can monitor tool joints as they go downhole, doubly assuring that all have the correct torque. Sometimes called an Iron Roughneck™, after the manufacturer of one such wrench.

I

IADC *abbr*: International Association of Drilling Contractors.

IBOP *abbr*: inside blowout preventer.

insert *n*: a removable, hard-steel, serrated piece that fits into the jaws of the tongs and firmly grips the body of the drill pipe or drill collars while the tongs are making up or breaking out the pipe. Also called die.

insert bowl *n*: a two-piece steel device with a tapered interior surface that fits into a one-piece or a hinged master bushing. It provides a place in the master bushing for crew members to set the slips. See also *master bushing, slips*.

inside blowout preventer (IBOP) *n*: a valve installed in the drill stem or in a top drive to prevent high-pressure fluids from flowing up the drill stem and into the atmosphere.

International Association of Drilling Contractors (IADC) *n*: an organization of drilling contractors headquartered in Houston, Texas. The organization sponsors or conducts research on education, accident prevention, drilling technology, and other matters of interest to drilling contractors and their employees.

Iron Roughneck™ *n*: manufacturer's trademark for a floor-mounted combination of a spinning wrench and a torque wrench. The Iron Roughneck eliminates the manual handling involved with suspended individual tools.

J

jaw *n*: see *tong jaw*.

jerk line *n*: a wire rope, one end of which is connected to the end of the breakout tongs and the other end of which is attached to the breakout cathead. When the driller activates the cathead, the cathead pulls on the jerk line with great force to apply torque to break out a tool joint (or to tighten a drill collar connection).

joint *n*: a single length (about 30 feet or 9 metres) of drill pipe or drill collar that has threaded connections at both ends. Two or three joints screwed together constitute a stand of pipe.

journal *n*: the part of a rotating shaft that turns in a bearing.

journal bearing *n*: a machine part in which a rotating shaft (a journal) revolves or slides.

K

kelly *n*: the heavy steel member, four- , or six-sided, suspended from the swivel through the rotary table and connected to the topmost joint of drill pipe to turn the drill stem as the rotary table turns. It has a bored passageway that permits fluid to be circulated into the drill stem and up the annulus, or vice versa.

kelly bushing *n*: a special device that, when fitted into the master bushing, transmits torque to the kelly and simultaneously permits vertical movement of the kelly to make hole. It may be shaped to fit the rotary opening or have pins for transmitting torque. Also called the drive bushing. See *kelly, master bushing*.

kelly bushing lock assembly *n*: a feature on four-pin kelly bushings installed on floating offshore rigs (which employ a conventional rotary table assembly) that secures the kelly bushing's pins to the master bushing's corresponding drive holes so that the kelly bushing will not separate from (lift off of) the master bushing as the rig heaves up and down with wind and wave motion. See *kelly bushing, master bushing*.

kelly cock *n*: a valve installed on one or both ends of the kelly. When a high-pressure backflow occurs inside the drill stem, the valve is closed to keep pressure off the swivel and rotary hose. See *lower kelly cock, upper kelly cock*.

kelly hose *n*: see *rotary hose*.

kelly saver sub *n*: a heavy and relatively short length of pipe that fits in the drill stem between the kelly and the drill pipe. The threads of the drill pipe mate with those of the sub, minimizing wear on the kelly.

kelly spinner *n*: a pneumatically operated device mounted on top of the kelly that, when actuated, causes the kelly to turn or spin. It is useful when the kelly or a joint of pipe attached to it must be spun up, that is, when the pipe must be rotated rapidly to make it up.

kick *n*: an entry of formation fluids into the wellbore. It occurs because the pressure exerted by the column of drilling fluid is not great enough to overcome the pressure exerted by the fluids in the formation drilled. If prompt action is not taken to control the kick or kill the well, a blowout will occur. See *blowout*.

L

lead tongs (pronounced "leed") *n pl*: when going into the hole, the tongs used to apply the necessary torque to make up a drill pipe tool joint to final tightness; when coming out of the hole, the tongs used to apply the necessary torque to break out a drill pipe tool joint. Compare *backup tongs*.

lifting sling *n*: an arrangement of special hooks that fit into receptacles in a master bushing or an insert bowl and chains or wire rope that are connected to a rig's air hoist, all of which enable crew members to insert or remove the master bushing and insert bowls as required.

link-tilt arms *n pl*: on a top drive, a device that, when actuated by the driller, tilts the unit's built-in elevators into a position to make it easy for crew members to latch the elevators onto a joint of pipe stored in the mousehole. See *top drive*.

load capacity *n*: the amount of weight a device can safely carry or support.

lock assembly *n*: see *kelly bushing lock assembly*.

locking device *n*: see *rotary locking device*.

long rotary slips *n pl*: slips designed to fit a square-drive master bushing, and whose taper length is 8 13/16 inches (224 millimetres). Compare *extra-long rotary slips*. See also *square-drive master bushing, slips*.

lower kelly cock *n*: a special valve normally installed below the kelly. Usually, the valve is open so that drilling fluid can flow out of the kelly and down the drill stem. When the mud pump is stopped, the valve can be manually closed with a special wrench to prevent pressurized fluids in the drill string from flowing into the kelly. Also called a drill stem safety valve, mud saver valve.

M

make a connection *v*: to attach a joint of drill pipe onto the drill stem suspended in the wellbore to permit deepening of the wellbore.

make a trip *v*: to hoist the drill stem out of the wellbore to perform one of a number of operations such as changing bits, taking a core, and so forth, and then to return the drill stem to the wellbore.

make up *v*: 1. to assemble and join parts to form a complete unit (as to make up a string of drill pipe). 2. to screw together two threaded pieces. Compare *break out*.

make up a joint *v*: to screw a length of pipe into another length of pipe.

makeup tongs *n pl*: tongs used for screwing one length of pipe into another, for making up a joint. Compare *breakout tongs*. See also *tongs*.

master bushing *n*: a device that fits into the rotary table to accommodate the slips and drive the kelly bushing so that the rotating motion of the rotary table can be transmitted to the kelly. Also called rotary bushing. See *kelly bushing, slips*.

mousehole *n*: an opening in the rig floor, usually lined with pipe, into which a length of drill pipe is placed temporarily for later connection to the drill string.

mud *n*: the liquid circulated through the wellbore during rotary drilling and workover operations. In addition to its function of bringing cuttings to the surface, drilling mud cools and lubricates the bit and the drill stem, protects against blowouts by holding back subsurface pressure, and deposits a mud cake on the wall of the borehole to prevent loss of fluids to the formation. Although it originally was a suspension of earth solids (especially clays) in water, the mud used in modern drilling operations is a more complex, three-phase mixture of liquids, reactive solids, and inert solids. The liquid phase may be fresh water, diesel oil, or crude oil and may contain one or more conditioners. See *drilling fluid.*

mud pump *n*: a large, reciprocating pump used to circulate the mud on a drilling rig. A typical mud pump is a single- or double-acting, two- or three-cylinder piston pump whose pistons travel in replaceable liners and are driven by a crankshaft actuated by an engine or motor. Also called a slush pump.

mud seal *n*: a synthetic rubber, ring-shaped washer that fits between parts of a device that are exposed to drilling mud and parts that need to be protected from drilling mud.

N

nozzle *n*: a restricted opening at the end of a passageway on a bit through which drilling fluid exits the bit. Many bits have three nozzles out of which the fluid jets with great velocity to keep the bottom of the hole clean. Nozzles come with openings of many sizes.

O

oil-bath reservoir *n*: in rotary table assemblies, a compartment in the base of the rotary table assembly that contains oil of a specified weight and viscosity and through which parts of the assembly move to be lubricated.

O-ring *n*: a special synthetic rubber ring that typically is used to form a seal around a cylindrical member in a piece of equipment.

P

packing *n*: a material used in a cylinder, in the stuffing box of a valve, or between flange joints to maintain a leak-proof seal.

pawl *n*: notches or slots machined into the table part of a rotary table assembly into which a bar on the rotary table assembly's locking device fits to keep the table from turning. See *rotary locking device, rotary table assembly.*

pin *n*: 1. the male threaded section of a tool joint. 2. on a bit, the threaded bit shank.

pin-drive master bushing *n*: a master bushing that has four drive holes corresponding to the four pins on the bottom of the pin drive kelly bushing.

pinion *n*: a gear with a small number of teeth designed to mesh with a larger wheel or rack.

pipe handler *n*: in a top drive, the power and spinning wrenches built into the unit that spins, makes up, breaks out, and backs up the pipe. See *top drive.*

power tongs *n pl*: see *power wrench*.

power wrench *n*: a set of tongs that is used to make up or break out drill pipe, tubing, or casing; the torque is provided by air or fluid pressure.

pressure-relief fitting *n*: a device that is set to open at a preset pressure to provide a point for excess pressure to exit from the device the fitting is mounted on. If any fluids are associated with the excess pressure, the fluids are usually routed to a receptacle, where they can be disposed of properly. On the drive shaft of a rotary table assembly, for example, relief fittings are provided to unload excess grease in the shaft.

prime mover *n*: an internal-combustion engine that is the source of power for a drilling rig.

psi *abbr*: pounds per square inch.

R

race *n*: a groove for the balls in a ball bearing or the rollers in a roller bearing.

rathole *n*: a hole in the rig floor, 30 to 36 feet (9 to 11 metres) deep, lined with casing that projects above the floor, into which the kelly and the swivel are placed when hoisting operations are in progress.

reaming *n*: a process in which the driller rotates and moves the drill stem up and down in the wellbore while circulating drilling fluid to clear the hole of debris or to prevent the drill stem from getting stuck.

relief fitting *n*: see *pressure-relief fitting*.

ring gear *n*: in a rotary table assembly, a circular, ring-shaped device with projections (teeth) that engage a beveled gear (a pinion) on the end of a drive shaft. The drive shaft is usually driven by a chain-and-sprocket arrangement from the drawworks. When engaged, the drive shaft turns the pinion, which meshes with the ring gear, to turn the rotary table.

roller *n*: on a kelly bushing, a cylindrical device that fits inside the bushing, whose exterior shape is matched to the kelly's shape so that they mate with the kelly when it is inside the kelly bushing. See *kelly bushing, roller assembly*.

roller assembly *n*: on a kelly bushing, an arrangement of rollers, roller pins, and roller bearings that mate with the kelly as it moves up or down inside the kelly bushing. The roller assembly transfers the turning motion of the kelly bushing to the kelly, and, at the same time, allows the kelly to move up and down freely. See *roller, roller bearing, roller pin*.

roller bearing *n*: a bearing in which a finely machined shaft (the journal) rotates in contact with a number of cylinders (rollers).

roller pin *n*: on a kelly bushing, a shaft that fits inside each roller on the bushing to affix the rollers to it. See *kelly bushing, roller assembly*.

rotary chain *n*: a chain drive powered by the drawworks that drives the rotary table assembly; it runs from the drawworks sprocket to a drive-shaft sprocket. The drive-shaft sprocket turns the drive-shaft assembly that drives the rotary table.

rotary helper *n*: a worker on a drilling or workover rig, subordinate to the driller, whose primary work station is on the rig floor. On rotary drilling rigs, there are at least two and usually three or more rotary helpers on each crew. Sometimes called floorhand, roughneck, or rig crewman.

rotary hose *n*: a steel-reinforced, flexible hose that is installed between the standpipe and the swivel or top drive. It conducts drilling mud from the standpipe to the swivel or top drive. Also called kelly hose.

rotary locking device *n*: on a rotary table, a steel pin, often spring-loaded, that, when engaged, fits into one of several notches machined onto the perimeter of the rotary assembly's turntable. When engaged, the lock prevents the turntable from turning.

rotary slips *n pl*: see *slips*.

rotary table *n*: the principal piece of equipment in the rotary table assembly; a turning device used to impart rotational power to the drill stem while permitting vertical movement of the pipe for rotary drilling. The master bushing fits inside the opening of the rotary table; it turns the kelly bushing, which permits vertical movement of the kelly while the stem is turning. See *kelly bushing, master bushing*.

rotary table assembly *n*: a rotating machine housed primarily inside a rectangular steel box with an opening in the middle for the kelly and the drill pipe that creates and transfers the turning motion for rotary drilling; parts of the assembly include the base, the rotary table, the master bushing, the drive-shaft assembly, the drawworks sprockets and drive-shaft sprockets, and the locking devices.

rotary table base *n*: a cast steel or reinforced fabricated steel shell that encloses the pinion end of the drive shaft and the rotary table.

rotary table locking device *n*: a small mechanical brake for the rotary table made of an iron rod and a notched wheel; its function is to stop the turning movement of the rotary table and hold it securely while the crew makes or breaks a connection or performs other jobs.

round trip *n*: the procedure of pulling out and subsequently running back into the hole a string of drill pipe. Making a round trip is also called tripping.

S

safety clamp *n*: see *wire-rope safety clamp*.

single *n*: a joint of drill pipe.

single-plane roller assembly *n*: in a kelly bushing, a single set of roller assemblies that are installed in the bushing at the same level, or plane. Compare *double-plane roller assembly*. See also *kelly bushing, roller assembly*.

slips *n pl*: wedge-shaped pieces of metal with serrated inserts (dies) or other gripping elements, such as serrated buttons, that suspend the drill pipe or drill collars in the master bushing of the rotary table when it is necessary to disconnect the drill stem from the kelly or from the top-drive unit's drive shaft. Rotary slips fit around the drill pipe and wedge against the master bushing to support the pipe. Drill collar slips fit around a drill collar and wedge against the master bushing to support the drill collar. Power slips are pneumatically or hydraulically actuated devices that allow the crew to dispense with the manual handling of slips when making a connection.

slip test *n*: a procedure to determine whether the slips' inserts (dies) are uniformly contacting the wall of the drill pipe. Paper is wrapped around the pipe and the slips are set. The paper is then unwrapped from the pipe and the pattern the slips made on the paper is examined. If uniform contact is not shown, the slips' inserts are replaced or the insert bowl is repaired or replaced.

snub line *n*: a strong wire rope attached to the end of the tongs and to one leg of the derrick to keep the tongs from turning too far when they are being used to make up, break out, or back up drill pipe or drill collars.

solid-body kelly bushing *n*: a kelly bushing that is cast in a single piece. Compare *split-body kelly bushing*. See also *kelly bushing*.

solid master bushing *n*: a master bushing made in one piece. Usually, solid master bushings have split insert bowls. Compare *hinged master bushing, split master bushing*. See also *insert bowl, master bushing*.

spinning chain *n*: a Y-shaped chain used to spin up (tighten) one joint of drill pipe into another. In use, one end of the chain is attached to the tongs, another end to the makeup cathead, and the third end is free. The free end is wrapped around the tool joint, and the cathead pulls the chain off the joint, causing the joint to spin (turn) rapidly and tighten up. After the chain is pulled off the joint, the tongs are secured in the same spot, and continued pull on the chain (and thus on the tongs) by the cathead makes up the joint to final tightness.

spinning wrench *n*: an air- or hydraulic-powered wrench used to spin drill pipe in making or breaking connections.

split-body kelly bushing *n*: a two-piece kelly bushing that is secured by hold-down nuts and bolts passing from the top of the bushing to the bottom of the bushing. Compare *solid-body kelly bushing*. See also *kelly bushing*.

split master bushing *n*: a master bushing that is made in two pieces. Each half has a tapered surface to accept the slips. Compare *hinged master bushing, solid master bushing*. See also *master bushing, slips*.

sprocket *n*: 1. a wheel with projections on the periphery to engage with the links of a chain. 2. a projection on the periphery of a wheel that engages the links of a chain.

square-drive kelly bushing *n*: a kelly bushing that has a square-shaped base that fits into a corresponding square-shaped recess in the master bushing. When the base is engaged in the recess, and the master bushing is rotated, the kelly bushing, the kelly, and the attached drill stem also rotate. See *kelly bushing, master bushing*.

square-drive master bushing *n*: a master bushing that has a square recess to accept and drive the square that is on the bottom of the square-drive kelly bushing. See *kelly bushing, master bushing, square-drive kelly bushing*.

square kelly *n*: a kelly with a four-sided (square) cross section. Compare *hexagonal kelly*.

stab *v*: to guide the end of a pipe into a tool joint when making up a connection. See *tool joint*.

stand *n*: the connected joints of pipe racked in the derrick or mast when making a trip. On a rig, the usual stand is 90 feet (27 metres) long (three lengths of pipe screwed together), or a thribble.

standpipe *n*: a length of pipe installed between the mud pump's output line and the rotary hose, through which drilling mud flows on its way to the swivel or top drive.

stem *n*: see *swivel stem*.

S-tube *n*: in a top drive, the S-shaped pipe that incorporates a gooseneck to which the rotary hose is attached. It conducts drilling mud into the top drive from the rotary hose. See *rotary hose, top drive*.

sub *n*: a short, threaded piece of pipe used to adapt parts of the drilling string that cannot otherwise be screwed together because of differences in thread size or design. A sub may also perform a special function. Lifting subs are used with drill collars to provide a shoulder to fit the drill pipe elevators. A kelly saver sub is placed between the drill pipe and the kelly to prevent excessive thread wear of the kelly and drill pipe threads. A bent sub is used when drilling a directional hole. "Sub" is short for "substitute."

sump *n*: a low place in a steel guard or casing that surrounds a moving chain or gear that requires constant lubrication. The sump holds a quantity of oil through which the moving parts travel and thus become lubricated.

swaging *n*: the tendency of a body, such as a length of drill pipe, to be bent by the action of a tool, such as the slips, applied to it with a great deal of force. In the case of drill pipe, if the slips are allowed to stop the drill stem as the driller lowers it into the hole, the slips can swage, or deform, the body of the pipe by the force they apply against the pipe when they settle into the master bushing.

swivel *n*: a rotary tool that is hung from the hook and traveling block to suspend and permit free rotation of the drill stem. It also provides a connection for the rotary hose and a passageway for the flow of drilling fluid into the drill stem.

swivel stem *n*: a length of pipe inside the swivel that is installed to the swivel's washpipe and to which the kelly (or a kelly accessory, such as the upper kelly cock) is attached. It conducts drilling mud from the washpipe and to the drill stem. See *washpipe*.

T

tail line *n*: see *guide line*.

tong arm *n*: the part of the tongs that extends behind the tong jaws and to which the jaws are attached. A snub line is attached to the end of the tong arm to prevent the tong from turning too far when it is being used to make up, break out, or back up pipe.

tong counterbalance *n*: a weight placed in the derrick or under the rig floor that is attached by means of wire rope to the tong hanger. Each set of tongs has a counterbalance to ensure that they hang at a convenient height above the rig floor.

tong hanger *n*: a relatively long narrow steel projection bolted to the tong arm to which wire rope is attached; suspends the tongs in the derrick.

tong jaw *n*: on a set of tongs, one of two hinged devices that crew members latch around elements of the drill stem to make up or break out such elements. Tong jaws normally have replaceable serrated inserts (dies) to grip the pipe.

tongs *n pl*: the large wrenches used for turning when making up or breaking out drill pipe, casing, tubing, or other pipe; variously called casing tongs, pipe tongs, and so forth, according to the specific use. Power tongs or power wrenches are pneumatically or hydraulically operated tools that serve to spin the pipe up tight and, in some instances, to apply the final makeup torque.

tool joint *n*: a heavy coupling element for drill pipe made of special alloy steel. Tool joints have coarse, tapered threads and seating shoulders designed to sustain the weight of the drill stem, withstand the strain of frequent coupling and uncoupling, and provide a leakproof seal. The male section of the joint, or the pin, is attached to one end of a length of drill pipe, and the female section, or the box, is attached to the other end.

top drive *n*: a device similar to a power swivel that is used in place of the rotary table to turn the drill stem. It also includes power tongs. Modern top drives combine the elevator, the tongs, the swivel, and the hook. Even though the rotary table assembly is not used to rotate the drill stem and bit, the top-drive system retains it to provide a place to set the slips to suspend the drill stem when drilling stops. Compare *conventional rotating system.*

torque *n*: the turning force that is applied to a shaft or other rotary mechanism to cause it to rotate or tend to do so. Torque is measured in foot-pounds, joules, newton metres, and so forth.

torque track *n*: on a portable top-drive unit, the steel rail that is mounted in the derrick and to which the top drive is attached. The top drive moves up or down on the track and, because it is firmly joined to the track, cannot turn as its drive shaft rotates the drill stem. Compare *guide rails, torque tube*. See also *top drive.*

torque tube *n*: a large cylindrical tube mounted in the derrick to which a top drive is attached and on which it can move up and down. Like guide rails, a torque tube prevents the top drive from turning when it is connected to the drill stem and is rotating it. Compare *guide rails*. See also *top drive.*

trip *n*: the operation of hoisting the drill stem from and returning it to the wellbore. *v*: shortened form of make a trip. See *make a trip.*

trip in *v*: to go in the hole.

trip out *v*: to come out of the hole.

trunnion *n*: metal pins protruding from the sides of a device that serve to mount the device onto another part.

turntable *n*: see *rotary table.*

two-piece insert bowl *n*: see *insert bowl.*

U

upper kelly cock *n*: a valve installed above the kelly that can be manually closed to protect the rotary hose from high pressure that may exist in the drill stem.

upset *v*: to forge the ends of tubular products, such as drill pipe, so that the pipe wall acquires extra thickness and strength near the end. Upsetting is usually performed to strengthen the pipe so that threads, or threaded pieces, such as tool joints, can be added to the pipe.

W

washpipe *n*: a short length of surface-hardened pipe that fits inside the swivel and serves as a conduit for drilling fluid through the swivel.

washpipe packing *n*: layers of dense flexible material that is stacked around the swivel's washpipe and the swivel's interior body seal. It helps seal between the static washpipe and the turning swivel body to ensure that high-pressure drilling mud flows through the swivel's stem and into the kelly.

wear sleeve *n*: a hollow cylindrical device installed around the swivel's rotating stem that absorbs the stem's rotation and helps form an oil seal between the stem and the oil-bath reservoirs inside the swivel.

wire-rope safety clamp *n*: a device that secures the end of a bend in the wire rope to the body of the wire rope. A typical safety clamp consists of a U-shaped bolt that is fastened to the wire rope with two nuts that thread onto the U-bolt. Depending on the diameter of the wire rope, which determines the amount of rope that is looped back on itself to secure the rope to the tongs or other device, clamps can vary in number from one to three.

Review Questions

LESSONS IN ROTARY DRILLING

Unit I, Lesson 4: Rotary, Kelly, Swivel, Tongs, and Top Drive

1. Name the three primary actions that rotary drilling performs downhole.

   (1) ______________________________

   (2) ______________________________

   (3) ______________________________

2. General components of the rotating system include, from top to bottom (fill in the blanks)—

   (1) ______________________________

   (2) Rotary (kelly) hose

   (3) Upper kelly cock

   (4) ______________________________

   (5) ______________________________

   (6) Kelly saver sub

   (7) ______________________________

3. What component of the rotary table assembly creates a turning motion?

   ______ A. the master bushing
   ______ B. the kelly
   ______ C. the rotary table
   ______ D. the turntable
   ______ E. the rotary table or the turntable

4. What piece of equipment frequently replaces the conventional rotating system on large offshore rigs? ______________________________

5. A rotary table assembly is—

   ______ A. a collection of many parts and gears
   ______ B. a rotating machine
   ______ C. often incorrectly called the rotary table
   ______ D. none of the above
   ______ E. A, B, and C

6. Name the two functions of the rotary table assembly.

   (1) ______________________________

   (2) ______________________________

7. The ______________________ is the principal piece of equipment in the rotary table assembly and consists of concentric circular gears, seals, and ball bearings.

8. The drive shaft is directly driven by the ____________________ and directly drives the ______________________.

   A. rotary chain, rotary table
   B. drawworks shaft sprocket, ring gear
   C. pinion gear, ring gear
   D. drawworks, kelly

9. Name the three ways that the master bushing may be designed or constructed.

   (1) ______________________________

   (2) ______________________________

   (3) ______________________________

10. What device engages the master bushing and permits the square or hexagonal kelly to move downward or upward through the rotary and its bushings?

    ____

11. What are the two functions of the master bushing?

    (1) ______________________________

    (2) ______________________________

12. Name the two types of drives for master bushings and kelly bushings.

    (1) ______________________________

    (2) ______________________________

13. The type, or design, of kelly bushing is determined by what two main features?

    (1) ______________________________

    (2) ______________________________

14. When and why are rotary slips used in drilling?

____

____

____

15. Describe the results of using slips that are too small or too large for the size pipe being handled.

(1) Smaller than the pipe: ______________________________

____

(2) Larger than the pipe: ______________________________

____

16. List three important guidelines to follow when using slips.

(1) ______________________________

(2) ______________________________

(3) ______________________________

17. What should be done when slip inserts (dies) or slip bodies become worn?

____

____

18. What is the purpose of performing a slip test?

____

____

19. Define the kelly.

____

____

____

20. The two primary jobs of the kelly are—

(1) ______________________________

(2) ______________________________

21. A kelly saver sub—
    ______ A. is a short threaded pipe that fits below the kelly
    ______ B. minimizes wear on the drill string
    ______ C. is a short threaded pipe that fits above the top of the drill string
    ______ D. all of the above

22. Name two devices used with the kelly to prevent drilling fluid from flowing up the drill stem and into the atmosphere.

    (1) ______________________________

    (2) ______________________________

23. Three functions of the swivel are—

    (1) ______________________________

    (2) ______________________________

    (3) ______________________________

24. Name four general maintenance practices for the swivel.

    (1) ______________________________

    (2) ______________________________

    (3) ______________________________

    (4) ______________________________

25. What are the two sets of tongs called when they are used (1) in making up a tool joint connection? (2) breaking out a tool joint connection?

    (1) ______________________________

    (2) ______________________________

26. Name four spinning or torquing power devices developed to take the place of spinning chain and tongs.

    (1) ______________________________

    (2) ______________________________

    (3) ______________________________

    (4) ______________________________

27. A top drive is—

_____ A. a part of a conventional rotating system
_____ B. a massive system of equipment that works much like a power swivel
_____ C. a massive system of equipment that replaces the rotary table and kelly
_____ D. none of the above

28. Name three advantages the top drive offers over conventional equipment:

(1) ______________________________

(2) ______________________________

(3) ______________________________

Answers to Review Questions

LESSONS IN ROTARY DRILLING

Unit I, Lesson 4: Rotary, Kelly, Swivel, Tongs, and Top Drive

1. (1) a downward movement of the drill stem
   (2) a turning action of the bit
   (3) the removal of cuttings from the hole
2. (1) swivel
   (4) kelly
   (5) lower kelly cock
   (7) rotary table assembly
3. E
4. a top drive
5. E
6. (1) during drilling, it rotates and transfers the turning motion to the kelly
   (2) when drilling is stopped, it holds or suspends the weight of the drill string
7. rotary table
8. A.
9. (1) solid
   (2) split
   (3) hinged
10. the kelly bushing
11. (1) it connects the rotary table to the kelly bushing and transfers rotation from one to the other
    (2) when drilling is stopped, the master bushing holds the slips
12. (1) pin drive
    (2) square drive
13. (1) the shape of the kelly it mates with
    (2) the number of roller planes it has
14. when drilling is stopped, to suspend the drill stem in the hole
15. (1) they damage pipe and the corner of the slips
    (2) they do not make contact with pipe all the way around the pipe; pipe can drop through the slips and damage them
16. (1) Never use slips to stop pipe when driller is lowering it.
    (2) Never allow the slips to ride the pipe while it is being pulled out of the hole.
    (3) Do not set the slips with the tool joint too high above the rotary table assembly.
17. discard the old slips and inserts (dies) and replace with new ones.

18. to detect wear on the slip inserts (dies) and the master bushing
19. a flat-sided, square or hexagonal, heavy steel pipe suspended from the swivel
20. (1) to transmit rotation to the drill string
    (2) to serve as a passageway for drilling mud
21. D
22. (1) upper kelly cock or valve
    (2) lower kelly cock (drill stem safety valve or mud saver valve)
23. (1) supports the weight of the drill stem during drilling
    (2) permits the drill stem to rotate
    (3) provides a passageway for the drilling mud to get into the drill stem
24. Any four of the following:
    (1) keep the swivel clean
    (2) coat the bail throat with grease
    (3) lubricate bail pins, oil seals upper bearing and packing
    (4) check oil level as recommended by manufacturer
    (5) check oil at intervals recommended by manufacturer
    (6) remove rust and apply weather protection as required
    (7) check and secure fasteners
25. (1) makeup and backup tongs (or lead tongs and backup tongs)
    (2) break out and backup tongs (or lead tongs and backup tongs)
26. (1) kelly spinner
    (2) spinning tongs
    (3) power tongs
    (4) hydraulic torque wrench
27. B
28. Any three of the following:
    (1) reduces the number of pipe connections and disconnections
    (2) cuts tripping in and tripping out time
    (3) eliminates the spinning chain and snub and jerk lines to manual tongs
    (4) allows drillers to ream long sections of the hole
    (5) facilitates well control